Zäzilia Seibold

Logical Time for Decentralized Control of Material Handling Systems

WISSENSCHAFTLICHE BERICHTE

Institut für Fördertechnik und Logistiksysteme
am Karlsruher Institut für Technologie (KIT)

BAND 89

Logical Time for Decentralized Control of Material Handling Systems

by
Zäzilia Seibold

KIT Scientific Publishing

Dissertation, Karlsruher Institut für Technologie (KIT)
Fakultät für Maschinenbau, 2016
Referenten: Prof. Dr.-Ing. Kai Furmans, Prof. Kevin Gue Ph.D.

Impressum

KIT Scientific Publishing

Karlsruher Institut für Technologie (KIT)
KIT Scientific Publishing
Straße am Forum 2
D-76131 Karlsruhe

KIT Scientific Publishing is a registered trademark of Karlsruhe Institute of Technology. Reprint using the book cover is not allowed.

www.ksp.kit.edu

Print on Demand 2016

ISSN 0171-2772
ISBN 978-3-7315-0567-9
DOI: 10.5445/KSP/1000057838

Logical Time for Decentralized Control of Material Handling Systems

Zur Erlangung des akademischen Grades eines

Doktors der Ingenieurwissenschaften

der Fakultät für Maschinenbau
des Karlsruher Instituts für Technologie (KIT)
genehmigte

Dissertation

von

Dipl.-Ing. Zäzilia Seibold

Tag der mündlichen Prüfung:	17. Juni 2016
Hauptreferent:	Prof. Dr.-Ing. Kai Furmans
Korreferent:	Prof. Kevin Gue Ph.D.

Danksagung

Die vorliegende Arbeit entstand während meiner Tätigkeit als wissenschaftliche Mitarbeiterin am Institut für Fördertechnik und Logistiksysteme des Karlsruher Instituts für Technologie. Ich möchte mich an dieser Stelle bei allen Personen bedanken, die zum Gelingen dieser Arbeit beigetragen haben.

Ich bedanke mich bei meinem Doktorvater Prof. Dr.-Ing. Kai Furmans für die vertrauensvolle Betreuung. Während meiner Zeit am IFL habe ich nicht nur im wissenschaftlichen Arbeiten viel von ihm gelernt. Prof. Kevin Gue PhD danke ich für die Übernahme des Koreferats. Als Betreuer meiner Diplomarbeit hat er die Begeisterung für wissenschaftliches Arbeiten und dezentrale Steuerungsalgorithmen in mir geweckt. Bei Prof. Dr.-Ing. Gisela Lanza bedanke ich mich für die Übernahme des Prüfungsvorsitzes.

Besonderer Dank gilt all meinen Kollegen für die gemeinsame Zeit am IFL. Allen Studierenden danke ich, die als Hiwis oder Abschlussarbeiter mit mir gemeinsam den GridSorter weiterentwickelt haben. Ebenso danke ich Gebhardt Fördertechnik und flexlog für die Unterstützung beim Aufbau und der Inbetriebnahme des GridSorter-Demonstrators.

Meinen Eltern und Geschwistern danke ich sehr herzlich für das Korrekturlesen und die interdisziplinären Diskussionen, die maßgeblich zur Verständlichkeit und Lesbarkeit meiner Arbeit beigetragen haben. Sie haben mich auf dem gesamten Lebensweg hierhin unterstützt und zu meiner Entwicklung beigetragen.

Aus vollem Herzen bedanke ich mich bei Andreas dafür, dass er so ist, wie er ist.

Karlsruhe, August 2016 Zäzilia Seibold

Kurzfassung

Die vierte industrielle Revolution zielt darauf ab, Produktionssysteme durch Verwendung zahlreicher kleiner elektronischer Bauelemente und Sensoren weiterzuentwickeln. Durch selbstständigen Informationsaustausch und daraus abgeleitete Entscheidungen versprechen diese Systeme mehr Flexibilität und Robustheit. Im Bereich der Fördertechnik hat die Forschung zu modularen, dezentral gesteuerten Systemen geführt. In den letzten fünf bis zehn Jahren wurden viele dezentrale Steuerungsalgorithmen für diese Systeme entwickelt, die in dieser Arbeit klassifiziert werden. Der erfolgreiche Umgang mit Deadlocks, also Situationen, in denen sich Ladungsträger endlos gegenseitig blockieren, wurde als Herausforderung erkannt. Das Deadlock-Risiko ist besonders hoch in Systemen mit dichten Transportweg-Strukturen und vielen Ladungsträgern. Bisher gibt es keinen allgemein anwendbaren Steuerungsalgorithmus für den Transport von Ladungsträgern, der Deadlocks strukturell verhindert und nicht nur während des Transports vermeidet. Der GridSorter, ein modulares Förder- und Sortiersystem mit gitterartiger Struktur, wurde als Anwendungsbeispiel für diese Arbeit gewählt, da er eine dichte Struktur mit vielen Ladungsträgern aufweist.

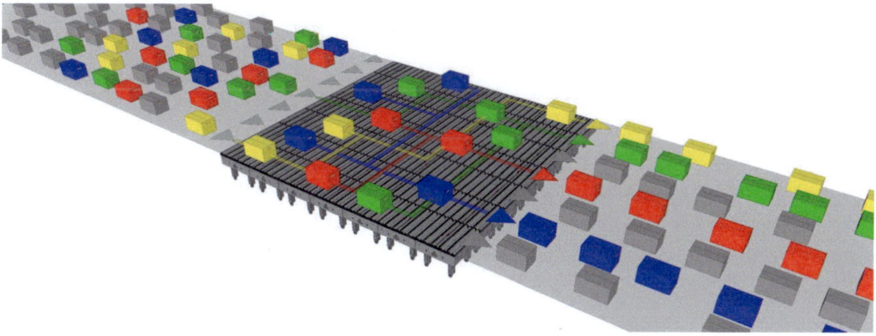

Schematische Darstellung eines exemplarischen GridSorter Systems

In dieser Arbeit wird ein neues Steuerungsprinzip für dezentral gesteuerte Materialflusssysteme vorgestellt: *Logische Zeit*, ein Prinzip für verteilte Systeme, das parallele Prozesse synchronisiert, wird auf dezentral gesteuerte Materialflusssysteme übertragen. Es wird bewiesen, dass das System dadurch Deadlock-frei ist. Darüber hinaus führt logische Zeit zu hoher Robustheit gegenüber schwankenden Transportzeiten, die z.B. durch Beschleunigungs- und Bremsvorgänge entstehen.

In dem dezentralen Steuerungsalgorithmus mit logischer Zeit für Materialflusssysteme besitzt jedes Modul einen Reservierungs- und einen Transportmanager. Der Reservierungsprozess kann als *Iterative Tiefensuche mit A*-Heuristik* beschrieben werden, die von der Gesamtheit der Reservierungsmanager aller Module für mehrere Ladungsträger parallel durchgeführt wird. Skalierbarkeit und Vollständigkeit des Reservierungsprozesses werden in dieser Arbeit behandelt.

Zur Untersuchung des Systemverhaltens mit dem vorgestellten Steuerungsalgorithmus wurde ein Simulationsmodell implementiert. Mithilfe einer Varianzanalyse wird der Einfluss mehrerer Steuerparameter untersucht: Wenn man die Größe des Suchbaums durch den gezielten Ausschluss alternativer Routen verringert, führt dies zu einer verbesserten Systemleistung bei gleichzeitig reduziertem Kommunikationsaufwand. Zwei Eigenschaften des Layouts beeinflussen den Durchsatz des Systems besonders: die Größe des Layouts und die Anzahl der Einschleusungen in Relation zur Größe des Layouts. Im Zusammenhang mit dem Prinzip der logischen Zeit werden verschiedene Beobachtungen gemacht: Bei geringer Systemlast entstehen Wartezeiten durch asynchrone logische Uhren, jedoch steigt die Synchronität der logischen Uhren mit zunehmender Systemlast. Eine zusätzliche Synchronisierung der einschleusenden Module verbessert zudem die Systemleistung.

Insgesamt wurde in dieser Arbeit ein dezentraler Steuerungsalgorithmus mit logischer Zeit für Materialflusssysteme entwickelt, die daraus resultierende Deadlock-Freiheit theoretisch bewiesen und die Systemleistung mithilfe eines Simulationsmodells untersucht. Außerdem wurde der Algorithmus erfolgreich auf einem 5×5-Demonstrator des GridSorters implementiert und getestet.

Abstract

The fourth industrial revolution aims to transform production systems by using numerous, small-scaled electronic devices. By sharing information and taking independent decisions, the systems promise more flexibility and robustness. In the field of material handling systems, research has led to modular systems with decentralized control. In the last five to ten years, many decentralized control algorithms have been developed for these systems that are classified in this thesis. Successful handling of deadlocks, i.e. situations in which transported items block each other infinitely, has been identified as challenge. The risk of deadlocks is particularly high in systems with dense structures and a high number of transported items. Until now, no generally applicable control algorithm exists for the transportation of boxes that structurally prevents deadlocks instead of avoiding them during transport. The GridSorter, a modular sorter with grid-like structure, is chosen as showcase system for this thesis because it forms a dense network with a high number of transported items.

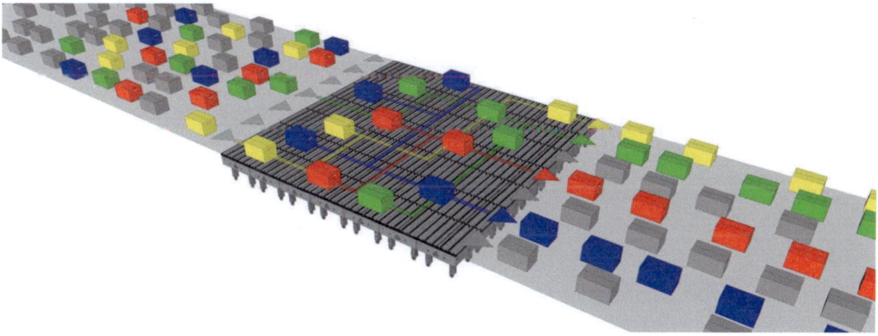

Schematic representation of an exemplary GridSorter system

In this thesis, a new control principle for decentralized material handling systems is presented: *Logical Time*, a control principle for distributed systems that synchronizes parallel processes, is transferred to decentralized control of material handling systems. With logical time, the material handling system is proven to be deadlock-free and is robust against varying transport times that arise for example from acceleration and deceleration.

In our decentralized control algorithm with logical time for material handling systems, each module owns a reservation and a transport manager. The reservation process can be described as *Iterative Deepening A** path search which is executed in parallel for several boxes by the collectivity of reservation managers of all modules. Scalability and completeness of the reservation process are discussed in this thesis.

To study system behavior with the presented control algorithm, a simulation model has been implemented. With the help of analysis of variance, the impact of several control parameters is investigated: Reducing the size of the search tree by limiting the number of alternative routes leads to better system performance and less communication effort. Two layout characteristics influence system throughput mainly: system size and the number of sources related to system size. Related to the principle of logical time, different observations are made: Under low system loads, waiting times emerge from asynchronous logical clocks but the synchronicity of logical clocks increases with increasing system load. An additional synchronization among source modules improves system performance.

In this thesis, a control algorithm with logical time for material handling systems with decentralized control has been developed, theoretically proven and system behavior has been studied in simulation. In addition, the control algorithm has been successfully implemented on a 5×5-demonstration system of GridSorter.

Contents

1 Introduction

A revolution is not a bed of roses. A revolution is a struggle between the future and the past.
-- Fidel Castro

Digitalization has changed many aspects of our lives: Thanks to electronic devices, we work, communicate and build relationships in a completely different way than ten to fifteen years ago. How does the use of electronic devices change production systems? *Industry 4.0* is the German buzzword to describe the transformation of production principles triggered by technological progress in the field of sensors, actuators, control devices and automated systems (Kagermann et al. 2012). It implies that we are standing at the beginning of another industrial revolution. *Cyber-Physical Systems* are part of this industrial revolution (Broy et al. 2012). The term stands for systems where the cyber-world merges with the physical world: the systems are able to sense the physical world and create a digital image of it. With this image, they assess alternative scenarios and interact accordingly with the physical world. At the same time, they communicate with other systems via networks.

This work looks at material handling systems as part of production systems. In this field, research has led to the development of modular material handling systems with decentralized control. Even though the development of these systems started before the term *Industry 4.0* was created, the system characteristics fit very well to the requirements of *Cyber-Physical Systems*: Each module is able to perform simple material handling tasks independently thanks to its sensors, actuators and its own control. By communicating with each other, several modules coordinate to fulfill complex system tasks.

What are the advantages of modular material handling systems with decentralized control? Seibold and Furmans (2016) have described some desirable characteristics of future material handling systems. We want to point out some of them: The absence of central control implies the absence of a single point of failure. The systems are designed in such a way that the system continues to work even though single modules have stopped working. This characteristic can be described as high robustness. Another big advantage is the facilitation of installation and reconfiguration. The system control configures itself, once the user has built up the physical system set-up. In addition, the production of these modules can be standardized and completed at the manufacturer's site. Also, we believe that decentralized control algorithms offer the chance that complex and varying material handling tasks can be realized with relatively simple control algorithms.

In the last five to ten years, much research has been conducted to develop decentralized control principles for material handling systems. The risk of deadlocks has been identified: A deadlock in a material handling system is usually characterized by items blocking each other on the way to their destination. The risk of deadlocks is particularly high in systems with dense structures and a high number of transported items: On grid-like patterns with high occupancy, for example, there are many different ways in which transported items could block each other. Different methods for deadlock handling have been developed for different systems. Nevertheless, especially in systems with high deadlock risk such as systems with grid-like patterns, the developed methods are quite complex and mostly specific to the task of the system. Besides, more and more material handling systems are being developed with grid-like patterns because they promise high efficiency per used surface.

The objective of this thesis is to develop a general control principle for decentralized material handling systems that prevents deadlocks and guarantees efficient system behavior even with disturbances leading to varying transport times. *Logical Time* is a control principle of distributed systems and has been used to synchronize parallel processes (Lamport 1978). Each process owns a *Logical Clock* that is only set forward if an event takes place. By assigning *timestamps* to events, a partial order is established that can be used for synchronization and negotiation. We believe that this principle can be transferred to decentralized control of material handling systems in order to prevent deadlocks effectively. In addition, the control principle leads to a high system robustness against transports times that differ from planned transport times.

We have chosen the GridSorter as showcase system to apply the developed algorithm because it forms a dense network with a high number of transported items. The GridSorter consists of rectangular conveyor modules forming a dense network. The GridSorter is primarily used for sorting of goods i.e. multiple items enter the system at sources and are transported to their specific destination. Figure 1.1 is a schematic representation of an exemplary system set-up: The boxes enter the system on one side and leave the system sorted by color on the other side.

Each of the modules is a conventional right-angle-transfer module and is able to convey a box in four directions with its sensors, actuators and its own control. By communicating with each other, the modules coordinate their actions. The GridSorter has no central control holding all information about the system state and taking decisions. In contrast, communication among the modules is required so that a module gets the necessary information to take decisions.

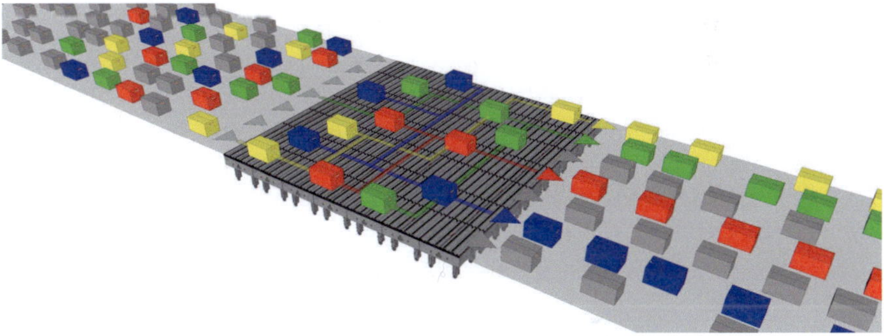

Figure 1.1: Schematic representation of an exemplary GridSorter system

The control algorithm must be able to fulfill different tasks. First, the conveyor modules must recognize the topology of the system because it is built up physically by the user and can be different each time. Second, the conveyor modules must take routing decisions in order to find routes to the destination of the goods. And third, the conveyor modules must perform the transport of the boxes without any collisions.

1.1 Problem Description and Research Questions

Many different material handling systems have been researched and developed. The handling of deadlocks has been identified as one key challenge (Mayer 2009; Krühn 2014) that must be resolved to guarantee system liveness i.e. functioning system behavior, especially in systems with grid-like patterns. System liveness includes the absence of deadlocks, livelocks and starvation. Resulting research questions to extract essential insights are:

> **1st Cluster of Questions** What systems have been developed in the last few years? How is their decentralized control designed? What kinds of deadlocks and livelocks are relevant in material handling systems? Which system characteristics increase the risk of deadlocks? How do other developers minimize the risk of deadlocks in their systems? How are deadlocks handled in systems of other research fields?

Most existing control algorithms are usually specific to one application. To simplify the algorithms for routing and deadlock handling, limitations are imposed. Logical time is a control principle that is applied in the field of distributed systems for the use of common resources. In this thesis, this principle will be applied to decentralized control of material handling systems. The objective is to develop a control principle that can be generally applied to different systems and that prevents deadlocks. If we want to transfer the principle of logical time to material handling systems, the principal questions are:

> **2nd Cluster of Questions** What are similarities between distributed systems and material handling systems? How can the principle of logical time be transferred to material handling systems? Is it possible to prevent deadlocks even in systems with high risk of deadlocks by using the principle of logical time?

If the principle of logical time is transferred to material handling, the existing control algorithms for route reservation and transport of boxes must be adapted or new algorithms must be developed. As exemplary system, we have chosen the GridSorter because it shows high risk of deadlocks and high routing complexity. Questions related to the control algorithm are:

> **3rd Cluster of Questions** How should the decentralized control be designed if the principle of logical time is applied? How is it possible to plan efficient routes for single boxes with acceptable duration and complexity? How can system scalability be achieved?

To study system behavior and performance, a simulation model is needed because the resulting behavior of decentralized decisions of conveyor modules cannot be analytically modeled (so far). Related questions are:

4th Cluster of Questions How can the system be modeled in simulation? What kind of simplifications are necessary to simulate the system? What input parameters influence the system behavior? What are key performance indicators?

Once the model is developed, it can be used to answer different questions related to the system behavior dependent on varying input parameters. The following questions summarize the intention:

5th Cluster of Questions How can system behavior be illustrated and interpreted? Which performance indicators are helpful? What kinds of interesting effects can be observed related to logical clocks? How do the different input parameters and layout characteristics influence system performance?

1.2 Structure of the Thesis

In **Chapter 2** an overview of existing material handling systems with decentralized control in research and industry is given. A special focus lies on algorithms for routing and deadlock handling. This is also the reason why an excursus is made to deadlocks, livelocks and starvation in operating systems and another excursus to routing in systems with autonomous guided vehicles. A classification of the systems follows, again focusing on the algorithms for routing and deadlock handling. At the end of the chapter, our choice of the GridSorter as showcase system is reasoned and the research gap is highlighted. The first cluster of questions is answered.

Chapter 3 is the main part of this thesis, and transfers the principle of logical time to material handling systems. After reasoning the choice of routing strategy, our approach to control the system with logical clocks is presented. It is shown how the order of transport steps is established and respected in order to prevent deadlocks. The chapter closes by proving the absence of deadlocks which also includes the absence of livelocks in our system. The second cluster of questions is answered.

In **Chapter 4**, the control architecture with all control components is introduced. The components responsible for the control of route reservation and transport are described in more detail because the principle of logical time is used in these components. Different methods to reduce the complexity of the route reservation are presented. The absence of starvation is shown by discussing the completeness of the reservation process. The third cluster of questions is answered.

Chapter 5 presents the simulation model of GridSorter. It discusses simplifications, and introduces input parameter and performance indicators. The fourth cluster of questions is answered.

In **Chapter 6**, system behavior is studied under different input parameters. This chapter consists of three parts: in the first part, the general system behavior is explained and the best combination of input parameters for the basic control algorithm is determined. The second part concentrates on the principle of logical clocks and interesting effects in their synchronicity. In the third part, the influence of the layout on the system behavior is studied, especially on throughput and communication effort. The influence of partial route reservation on these indicators is included. The last cluster of questions is answered in this chapter.

After responding to all clusters of questions, the final chapter draws an overall conclusion to the thesis. Have all research questions been answered? How could the principle of logical time be applied to other systems?

2 Material Handling Systems with Decentralized Control

Curiosity is always the first step to resolving a problem.
-- Galileo Galilei

In section 2.2, we introduce projects in research and industry in the field of material handling systems with decentralized control. The keywords used to describe their properties are quite different: *Plug&Play*, *Grid...*, *Internet of Things*, *Physical Internet*, *cognitive*, *cellular* ... etc., but all systems have the same key idea. The decisions are not taken by one central control, but the system is divided into modules that take decisions (more or less) self-dependently. The objective is to increase flexibility by reducing the effort for installation or reconfiguration of a system. Furmans et al. (2010) describe desired properties of future material handling systems and propose suitable design patterns to achieve the properties. Seibold et al. (2015) reappraise the propositions and study how the development of recent years has brought the ideas forward.

All material handling systems with decentralized control face similar challenges in matters of control: The routing strategy is essential for system performance and for the handling of deadlocks. This is why section 2.3.2 devotes attention to strategies for routing and deadlock handling in existing systems. We also include an excursus to systems of autonomous guided vehicles (AGV) with centralized control because they have developed interesting approaches using time-window-based routing.

In section 2.4, we classify the presented systems according to criteria mainly related to routing and deadlock handling strategy and draw conclusions on research gaps.

2.1 A Graph-Based System Description

We introduce a graph-based system description as an aid to understanding the key difference in the control of continuous and discontinuous material handling systems. Continuous material handling systems are able to generate a continuous flow of goods by means of conveyors. The possible transport routes are defined by the installed conveyor equipment. Discontinuous material handling systems transport the goods with autonomous guided vehicles (AGVs), each of them being able to transport a certain number of goods. For discontinuous systems, we only consider track-guided systems because they have clearly defined transport routes like continuous conveyor systems. This common characteristic makes it possible to compare the control principles of both system types. Table 2.1 shows the common characteristics and the differences between a conveyor system and an AGV system as examples for continuous and discontinuous material handling systems respectively.

Some common characteristics exist that become clear if the systems are modeled as graphs (see Table 2.1):

- In both system types, there are **mobile objects** which must be moved from a starting position to a destination. The mobile objects need resources to remain in one position or to move to another position.
- The physical resources form a network that can be described as a **graph with nodes and edges**. A node represents the possibility for a mobile object to remain in one position. An edge connects two nodes and enables a mobile object to move from one node to another. If nodes are connected with more than one other node, a routing decision has to be taken in which direction a mobile object should move.

Nevertheless, there are differences in the control of the systems (see Table 2.1):

- **Which are the active objects?** Within continuous systems, the nodes i.e. the resources are active, meaning they are equipped with actuators, sensors and eventually a decentralized control. Within discontinuous systems, the mobile objects i.e. the vehicles are active, and are equipped with actuators, sensors and a control.
- **How is the destination of a mobile object assigned?** Within continuous systems, each mobile object has its specific destination location (or none if the object should remain somewhere in the system for

storage purpose). Within discontinuous systems, a mobile object inherits its destination from the goods it transports. Depending on the assignment of transport orders to mobile objects, the mobile objects have to move to different destinations.

With decentralized control, the two system types differ in terms of routing decisions: Within continuous systems, the nodes i.e. the resources are active and thus able to take routing decisions[1]. Within discontinuous systems, it is the mobile object that should be able to take routing decisions.

Common Characteristics	**Continuous MHS** e.g. conveyors	**Discontinuous MHS** e.g. AGVs
Mobile objects	boxes	vehicles
Nodes and edges defined by	conveyors	lanes
Differences		
Active Objects	nodes i.e. conveyors	mobile objects i.e. vehicles
Destination of mobile object	specific	interchangeable

Table 2.1: Common characteristics and differences of continuous and discontinuous material handling systems (MHS) with decentralized control

This theoretical system description can help to classify the systems described in section 2.2. It is also relevant for the understanding of deadlock risk and the control algorithms and is resumed in section 2.3.1 and Chapter 3.

2.2 Existing Systems in Research and Industry

In this section, we introduce seven continuous material handling systems featuring decentralized control. We first present the "FlexConveyor" because this system is the basis for several other systems including the GridSorter. We then present four track-guided, discontinuous material hand-

[1] Some research projects aim at turning the boxes in active objects being able to take routing decisions (Emmerich et al. 2012). These systems offer the opportunity to rethink the existing control algorithms.

ling systems featuring decentralized control. We only present systems with track-guided vehicles and exclude systems with free-moving vehicles from this overview because they face different challenges in routing and deadlock handling. All systems with the term "Grid" have the common idea that goods are stored and transported in a grid pattern generated by conveyor modules or routes of AGVs.

After presenting the systems with decentralized control, we give an excursus presenting a couple of systems with centralized control showing the similarity of grid-like movement of boxes. For these systems, no details of the control algorithms have been published. Nevertheless, it shows that control algorithms for grid-like systems are in great demand.

FlexConveyor is a modular material handling system with decentralized control (see Figure 2.1). According to Mayer (2009), each module is identical: a rectangular right-angle-transfer which is able to transport goods in the four cardinal directions. Nowadays, different kinds of modules can be combined (flexlog GmbH 2015). The conveying system can be easily built by the user by combining the conveying modules, because they are equipped with wheels and can be connected without any tools. The connection between neighbors is physical, electronic and electrical. Each module has its own control and communicates with its neighbors to take decisions. For each box entering the system, a route is reserved from source to destination to prevent opposing routes on bidirectional conveying modules. With so-called deadlock tokens, loops of boxes waiting for each other are avoided (Mayer and Furmans 2010).

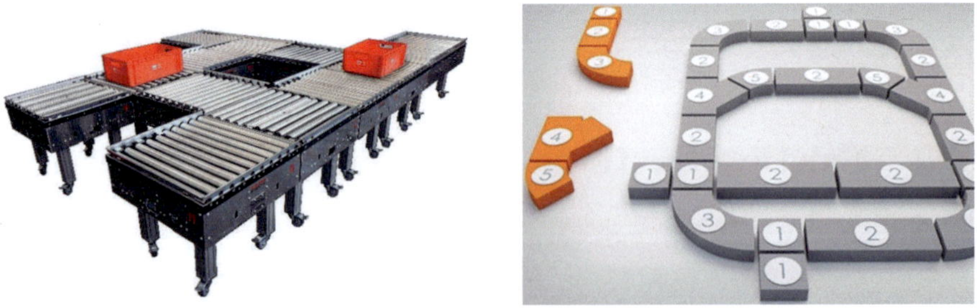

Figure 2.1: FlexConveyor, Left: photograph of identical, transfer modules (Seibold et al. 2013), Right: graphic of a network built of different module types (flexlog GmbH 2015)

GridSorter is based on the idea of FlexConveyor. Again, the system is built of rectangular transfer modules (as in the early version of FlexConveyor), each being able to communicate with its neighbors and to take decisions (Seibold et al. 2013). The difference to FlexConveyor is the system task and the density of the conveying network. The main task of GridSorter is to sort goods, i.e. to transport goods to different destinations based on case-specific, known criteria. In order to reach a high module- and space-efficiency, the conveying network is built as densely as possible: the result is a grid-like network topology (Seibold et al. 2014). The topology can have any form and the sources and destinations can be at any side module of the conveying network. Seibold et al. (2013) adapt the routing and deadlock handling of FlexConveyor to GridSorter. With this control, deadlocks occur if high throughput is required and if order input is not controlled accordingly.

GridStore and FlexConveyor are based on the same idea. In GridStore, a dense network transfer modules with decentralized control, the grid, is used to store goods. GridStore aims to combine the apparently opposing objectives of high throughput and high density of a storage system (Gue et al. 2014). Items are retrieved on one side of the grid and replenished on the opposing side. The used control algorithm is called "Virtual Aisles" because stored items are moved out of the way of requested items in order to form aisles. The decision as to how items should be moved is taken by each module stepwise based on the current situation.

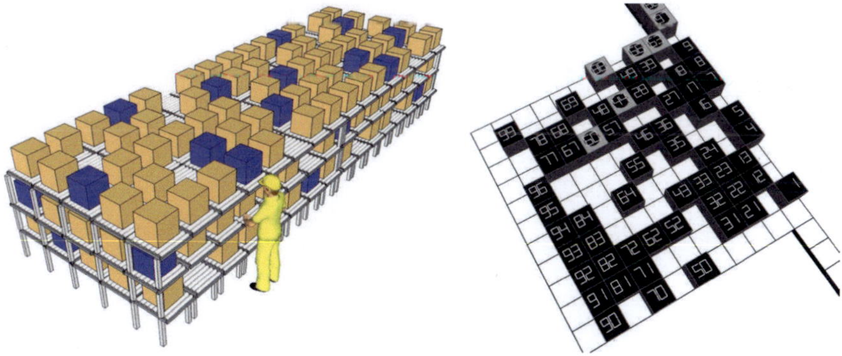

Figure 2.2: Left: Graphic of GridPick (Uludag 2014).
Right: Graphic of GridSequence (Gue et al. 2012)

GridPick is a special implementation of GridStore. Items are retrieved and replenished on the same side of the grid in order to supply items of one order to a picker (see Figure 2.2). In comparison to commonly used flow racks, the walking distance of a picker and therefore the pick time of one order can be reduced. To enable movements in all four cardinal directions, the "Virtual Aisles" algorithm has been adapted and balances the number of items per row (Uludag 2014).

GridSequence uses like GridStore and GridPick a dense network of transfer modules with decentralized control. The system task is to establish a certain sequence within the retrieved items (see Figure 2.2). The "Virtual Aisles" algorithm has been adapted by assigning an intermediate destination to each item depending on its sequence number (Gue et al. 2012).

Cognitive Conveyors are built of small-scaled modules with decentralized control (see Figure 2.3). Since one module is as small as one roller, several modules must form groups in order to transport items: Omnidirectional movement of items is possible. The main system task is to transport items, but it is also possible to use the system for buffering and sequencing (Krühn et al. 2013). Krühn (2014) describes the control algorithm which reserves a route from source to destination for each box and avoids deadlocks during transport.

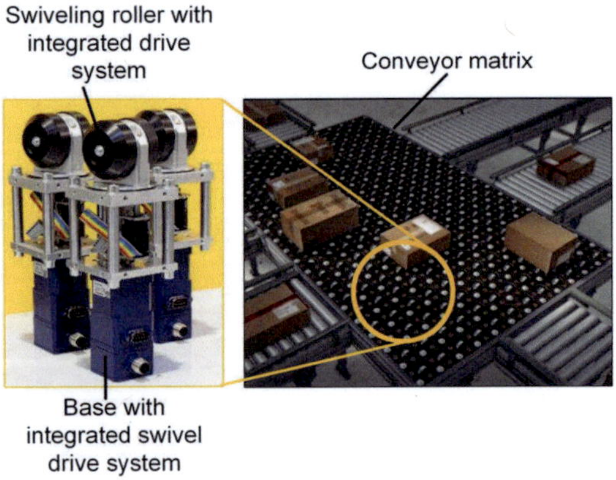

Figure 2.3: Graphic of Cognitive Conveyors (Krühn 2014)

Conveyor Showcase for the Internet of Things Within the "Internet of Things", objects communicate with each other and take decisions self-dependently (ten Hompel 2010). The control of continuous material handling systems is achieved by agents who are responsible for certain objects like single conveyor modules or the load being transported (Feldhorst et al. 2010). The system has been industrialized under the name THINGtelligence (Lanfer Automation GmbH & Co. KG 2015). Roidl (2012) describes a control algorithm where multiple routes are planned from each source to each destination with a heuristic based on the path finding algorithm of ants. Thanks to this heuristic, load-dependent routing is possible. Each conveyor module receiving an item decides probabilistically in which direction the item should be transported based on the routing information. Another similar control algorithm is described by Hofmeister et al. (2010). Again, routes are planned from each source to each destination based on topology and load information. The difference is that the routing information is collected and cumulated for certain time windows to capture the change of load with time which is then considered in the route planning. The control algorithm has been tested in different scenarios (Elger et al. 2010). Both control algorithms are only valid in unidirectional conveyor networks.

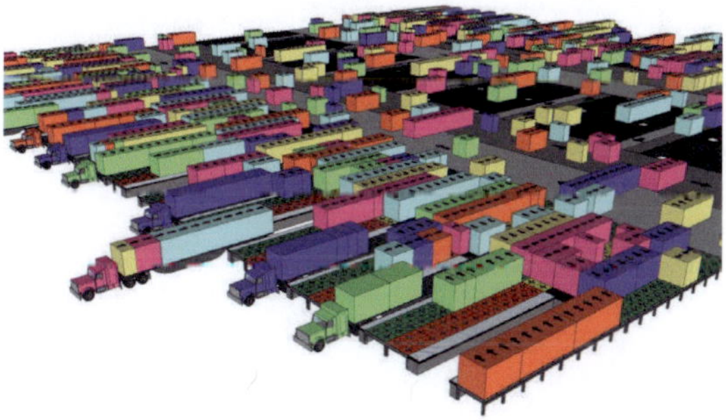

Figure 2.4: Schematic 3D-model of the modular warehouse (Montreuil et al. 2012)

Modular Warehouse as Part of the Physical Internet The Physical Internet stands for the vision in which standardized loads and material handling systems facilitate global transport tasks. One solution within the Physical Internet is the Modular Warehouse (see Figure 2.4) where stored objects are able to perform rectilinear movement with specific material handling devices (Sittivijan 2014). Even though the system in Figure 2.4 seems to be a continuous material handling system, the control algorithm is described in the same way as for a discontinuous one because the objects take decisions. We therefore consider this system as a discontinuous material handling system where each object is carried by a vehicle. Objects are active if they need to be transported to a specific destination. The route is planned on the basis of the information where other objects are located and if they have planned to move away. When executing the transport along the planned path, conflicts are resolved by using unique priorities of the objects. An object with lower priority is forced to move from its planned route if it is required to give way to an object with higher priority. Deadlocks are detected with a centralized control algorithm and resolved by changing the priorities of the objects. Some specific livelocks cannot be resolved even though they are detected by the central control.

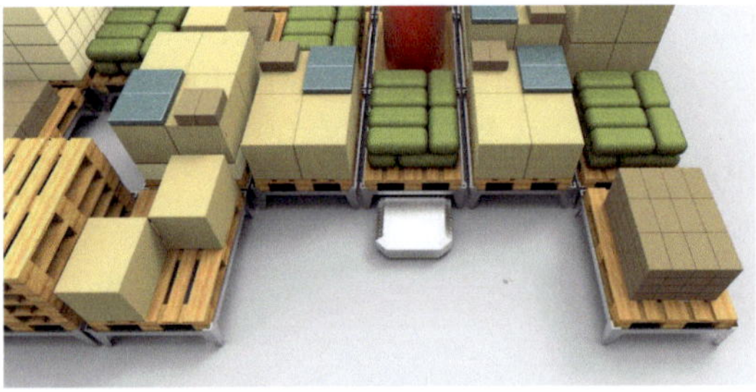

Figure 2.5: Graphic of GridFlow (Schwab 2015)

GridFlow with Transport Vehicles is a system where multiple transport vehicles enable the movement of pallets that are stored very densely (see Figure 2.5). The pallets are placed in a grid pattern and the vehicles move underneath the pallets following this grid pattern. Like within GridStore, the objective is to achieve high storage density while guaranteeing low retrieval times. The system features decentralized control, i.e. each vehicle decides autonomously. Schwab (2015) presents a control algorithm where each vehicles decides on its next movement based on communication with all other vehicles. Deadlocks and livelocks are detected and resolved by ordering the vehicles using unique priorities. The vehicle with lower priority is forced to give way to the vehicle with higher priority.

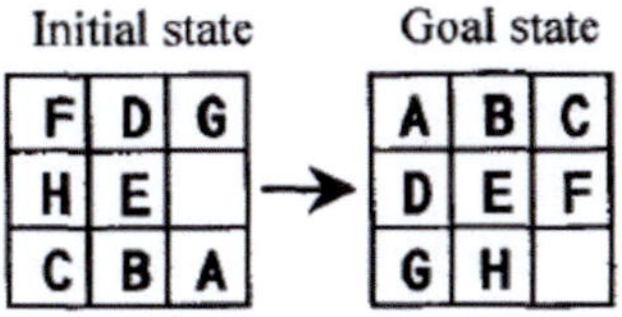

Figure 2.6: Example for a warehouse transport problem (Hama et al. 2002)

Distributed Agents in a Cellular Warehouse Hama et al. (2002) describe a cellular warehouse where transport agents are able to transport items in a grid-like pattern. The system is brought from a current state to a target state by the agents negotiating with each other. Each agent chooses its action, i.e. movement, based on a built-in behavior function given by artificial neural networks. Whether or not an agent is allowed to perform its action is decided by negotiation with the surrounding agents to establish an ordering. With this control mechanism, the agents are able to solve transport problems similar to the 8-puzzle-problem (see Figure 2.6).

Figure 2.7: Photograph of electric monorail system (Chisu et al. 2010)

Agent-Based-Controlled Electric Monorail System Chisu et al. (2010) describe the modularization of an electric monorail system (see Figure 2.7) and its surrounding conveyor modules. Each resource, for example the vehicles and the conveyor modules are controlled by an agent. Routes are reserved in advance because the tracks can be used bidirectionally. Each vehicle calculates the shortest path and communicates its decision to the other agents for reservation of the chosen route (Wilke 2006). The agents communicate with each other by means of a blackboard in order to reduce the number of message transfers (Tenerowicz-Wirth 2013).

Excursus: Systems with Grid-like Patterns and Centralized Control
For all of the three following systems, no details of the control algorithm have been published. Nevertheless, they stand out due to the mechanical realization of the grid-like architecture.

The *Celluveyor* presented by Uriarte et al. (2015) consists of hexagonal modules able to transport boxes unidirectionally thanks to so-called omni-wheels (see left side of Figure 2.8). The resulting functionalities are similar to those of the Cognitive Conveyor.

Figure 2.8: Left: Photograph of Celluveyor (Uriarte et al. 2015)
Right: Schematic representation of the shuttle system Store Biter (Gebhardt Fördertechnik GmbH 2015)

Yalcin et al. (2015) investigate the suitability of grid-based storage systems for early baggage in airports. They have chosen the approach of a centralized control. Details have not been published yet. The baggage is transported by electromagnetic power. The system is called *Fluid Logistics*.

The *Motion Cube* uses a nonslip traction to move specialized load carriers in production (Festo AG & Co. KG 2015). The movement principle is also described by Tadakuma et al. (2012a) and Tadakuma et al. (2012b): Gear wheels are positioned at 90° to each other and are able to realize linear movement of the carriers in all directions (see Figure 2.9). The advantage of nonslip traction is the highly precise positioning of the carriers.

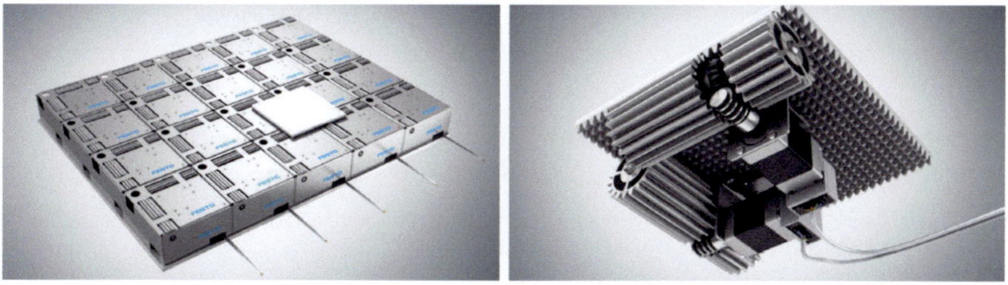

Figure 2.9: Schematic representation of Motion Cube
(Festo AG & Co. KG 2015)

The trend to grid-like pattern is not only observable in continuous material handling systems but also in discontinuous ones. In a shuttle storage system developed by Gebhardt Fördertechnik GmbH (2015), vehicles are enabled to move in up to three dimensions (see right side of Figure 2.8). Just like in continuous systems, a high network density increases the risk of deadlocks and the level of routing complexity.

2.3 Strategies for Routing and Deadlock Handling

Past and ongoing research on decentralized control of material handling systems has shown that deadlock handling is a critical success factor. In some of the previously presented systems, very complex deadlock handling mechanisms have been developed and implemented. The objective of all these systems is to define a decentralized control algorithm that guarantees a functioning and efficient system behavior while causing low communication effort. Functioning system behavior is given if every box is eventually transported to its destination. In the following, this property is called system liveness. System efficiency is usually measured by system throughput (sometimes as a function of system density). The communication effort can be measured by the number of sent messages per reservation or per time unit.

Deadlock handling is a basic requirement for every operating system and has been widely studied in the field of resource allocation. In section 2.3.1, we introduce the conditions under which a deadlock arises and the strategies for dealing with resource deadlocks in operating systems. In systems with decentralized control, dealing with deadlocks becomes even harder because there is no central control instance that has all information about current states of resources and processes.

In section 2.3.2, we examine how existing routing strategies of material handling systems with decentralized handle deadlocks. We also discuss how these routing strategies fit to the deadlock handling strategies used in operating systems.

In section 2.3.3, we present some research from the field of multi-agent routing. Time-window-based routing has already been applied to systems of multiple moving objects, for example AGVs. Since some challenges such as deadlock handling are similar, the results from this research could be helpful for material handling systems with decentralized control.

2.3.1 Resource Deadlocks, Livelocks and Starvation

Tanenbaum and Bos (2015, p. 439) define a deadlock as follows:

> *A set of processes is deadlocked when every process in the set is waiting for a resource that must be released by another process in the set.*

Coffman et al. (1971) define four conditions that must coexist for there to be a deadlock. A deadlock can only occur if all four conditions are fulfilled. If one of the conditions can be excluded, the risk of deadlocks is barred. Tanenbaum and Bos (2015, p. 440) describe these conditions as follows:

Mutual Exclusion Each resource is either currently assigned to exactly one process or is available.

Hold and Wait Processes currently holding resources that were granted earlier can request new resources.

No Preemption Resources previously granted cannot be forcibly taken away from a process. They must be explicitly released by the process holding them.

Circular Wait There must be a circular chain of two or more processes, each of which is waiting for a resource held by the next member of the chain.

Figure 2.10 shows two examples of deadlocks that could occur in a modular conveying system. The relevant modules are occupied with boxes each waiting for the next conveying module to become available. Mayer (2009) transfers the four deadlock conditions defined by Coffman et al. (1971) to the decentralized control of FlexConveyor. He states that the first three conditions are always present:

- Mutual exclusion is given because a conveyor module can only carry one box.
- The Hold-and-Wait condition is given because a box remains on a conveyor until the next conveyor on its route becomes available.
- No preemption is possible, as the box cannot be taken from the conveyor.

Consequently, the circular wait condition is the only condition that can be attacked. In Figure 2.10, two kinds of deadlocks are shown where boxes are waiting in a circular chain for each other. On the left, opposing routes have been chosen for the two boxes. On the right, the routes of the boxes are not opposing but forming a loop. These conditions can also be formulated for discontinuous material handling systems.

Not only deadlocks hinder processes from their termination, but livelocks and starvation also exist as closely related problems (Tanenbaum and Bos 2015). A livelock is described as a situation with several processes repeatedly switching between resources without progressing. Starvation can be described as a situation where the rule of resource allocation repeatedly disadvantages a certain process. Consequently, the requested resource is never requested to the one process which is why the process will never terminate.

In a material handling system, a livelock is characterized by an object performing the same cycle of movements all over again (Schwab 2015) without being transported to its destination. Starvation could happen at a crossing where the objects from different directions compete for a resource. If one direction has high throughput and is always prioritized, the objects from the other direction will never get closer to their destination.

Figure 2.10: Deadlock because of opposing routes (left) and because of boxes waiting in a loop (right)

Tanenbaum and Bos (2015, p. 343) describe four strategies for dealing with deadlocks:

Just ignore the problem: Maybe if you ignore it, it will ignore you.

Detection and recovery: Let deadlocks occur, detect them, and take action.

Dynamic avoidance by careful resource allocation.

Prevention by structurally negating one of the four required conditions.

Ignoring the problem of deadlocks is only a solution if the occurrence probability of deadlocks is low and if the occurrence is accepted by the system user. For deadlock detection, different algorithms exist. Recovery could be achieved by preemption of resources or the rollback or killing of processes. The objective of deadlock avoidance is to keep the system in safe states where no deadlocks can occur. Before a resource is allocated to a process, it must be checked that this results in a safe system state. For deadlock prevention, the system must be designed in such a way that at least one of the four conditions is negated. The ordering of resources is one method for deadlock prevention. By defining an order in which the resources must be requested, the circular wait condition is structurally negated. According to Raynal (2013), the drawback of this method is its inefficiency because long chains of waiting resources can build up. Lynch (1981) proposes defining only partial ordering between resources in order to reduce waiting chains. She uses the method of vertex coloring as an approach to find a good partial ordering for a set of resources knowing which resources are needed by each process.

Which of the four strategies can be applied in material handling systems? The first, ignoring the problem, is only acceptable if the occurrence of deadlocks is improbable because of system design. Otherwise, ignoring deadlocks is not acceptable because each box must reach its destination. Detection of deadlocks and system recovery by rollback is possible, but this strategy would have a negative impact on throughput because previously performed transports would have to be performed backwards again. Also, these rollback strategies could become complex because of potential livelocks occurring. Within deadlock avoidance, a compromise has to be found between careful resource allocation and system performance. Additionally, deadlock avoidance must take place during the transportation of boxes and may cause delays if the computing time is high. In material handling systems with decentralized control, a deadlock prevention strategy that does not decrease system performance is preferable.

The differentiation between these four strategies for deadlock handling is controversially discussed. Levine (2005) refutes the distinction between avoidance and prevention because both strategies aim to negate one of the four deadlock conditions. Observing material handling systems, one could even say that deadlock detection and recovery corresponds to prevention of deadlocks: Boxes never wait in a loop for each other because the control directly reacts once a loop is detected. Consequently, the waiting of boxes is not visible for an external observer. Nevertheless, the distinction between the four strategies seems to be natural.

In the context of material handling system, we describe the four strategies as follows.

Ignoring Deadlocks are ignored because they are unlikely to happen.

Detection & recovery At some (very) short moment, the loop of waiting objects exists. For example, if two conveyor modules decide independently from each other to send a box, detect an opposing deadlock and revise the previous decision.

Avoidance The deadlock of objects in a loop is detected shortly before it happens. The occurrence is avoided by waiting of objects.

Prevention The deadlock of objects in a loop is prevented long before it could happen. The system structure is changed.

Examples for each of the four strategies will be presented in the following section, in which we discuss existing routing strategies.

2.3.2 Routing in Material Handling Systems with Decentralized Control

Let us first explain what the term *routing* stands for before describing what different kinds of routing strategies exist in material handling systems with decentralized control. The term *routing* is differently defined depending on the field of application. In addition, it is used differently by different authors from identical fields. We refer to the definition of *routing* within communication networks and systems of autonomous guided vehicles. In communication networks, data packages are sent from source to destination through a network of routers which has similarities to sending physical boxes through a network of conveyors. In AGV systems, single vehicles move self-dependently through a network of passive resources.

In communication networks, "the network layer is routing packets from the source machine to the destination machine" (Tanenbaum 2011, p. 380). This definition implies that the routing contains all related tasks necessary to send a package from source to destination, including the path finding and the actual sending of the packages. Tanenbaum (2011, p. 49) differentiates between the routing algorithm and the forwarding algorithm as follows:

> How the network makes the decision as to which path to use is called the **routing algorithm**. [...] How each router makes the decision as to where to send a packet next is called the **forwarding algorithm**.

Usually, packages are sent from source to destination via multiple routers. In each router, two different processes are responsible for routing and forwarding (Tanenbaum 2011, p. 381): The routing process "is responsible for filling in and updating the routing tables" and the forwarding process "handles each packet as it arrives, looking up the outgoing line to use for it in the routing tables." In systems with virtual circuits, the routing algorithm is not started for each package individually.

However, in the field of the control of Autonomous Guided Vehicles (AGVs) *routing* describes the planning of a route for a specific vehicle and the execution of the route (Qiu et al. 2002). In this thesis, we use the term *routing* according to Qiu et al. (2002): *Routing* comprises the route planning for a specific box and the execution of the transport.

In material handling systems, the routing algorithm directly influences the deadlock handling strategy because it controls the possibility of forming a circular chain of waiting boxes. We now present three criteria that are helpful to describe different routing strategies (see Figure 2.11):

- Route planning can be done for each individual object or for several objects moving to the same destination. Route planning for a group of objects corresponds to the principle of *forwarding* packages in virtual circuits in communication networks (Tanenbaum 2011).
- Route planning and execution can be decoupled in time, i.e. planning is done before execution is started. Or route planning can be performed during execution. Qiu et al. (2002) differentiates between *static* and *dynamic* routing.

- The route planning can be performed *time-independent* or *time-window-based* by considering future system states. These two approaches are called different names by many authors.

Based on these three criteria, it is possible to describe eight different routing strategies. As shown in Figure 2.11, we group them into four mutual exclusive routing strategies. This classification is also a combination of classifications by Mayer (2009), Qiu et al. (2002), ter Mors (2009) and has been developed together with Geier (2015).

Forwarding Routes are planned from each source to each destination for a group of objects. During route execution, objects move to the next resource based on this information.

Dynamic Routing For each individual object, the routing decision i.e. where to move next is taken dynamically taken during route execution.

Time-independent Route Reservation The route from source to destination is reserved for each individual object before it moves. Route planning is decoupled from execution. The reservation is kept active until the object has passed.

Time-window-based Route Reservation Again, the route from source to destination is reserved for each individual object before it moves. The resources are only reserved for the duration of the time window the object is planned to need them.

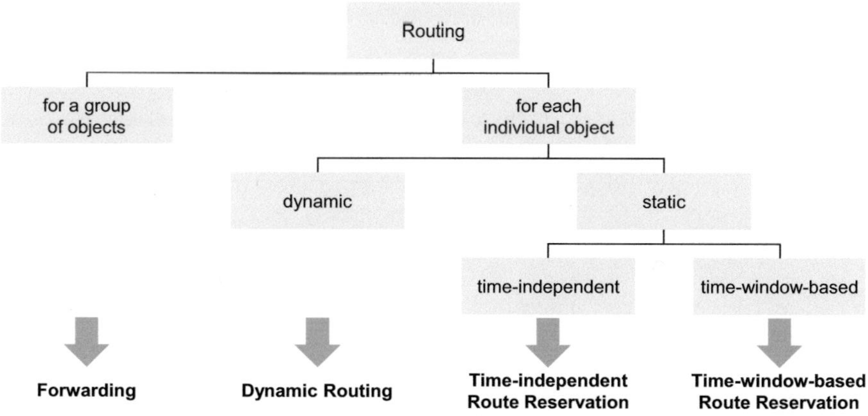

Figure 2.11: Criteria for routing and four resulting routing strategies

We will now describe how the four routing strategies are implemented in existing systems and what kind of deadlock handling strategy is used in combination.

Forwarding is used in unidirectional conveying networks within the vision of the Internet of Things. During route planning, all possible routes from source to destination are rated, usually based on distance and load. These possible routes correspond to virtual circuits in communication networks. In addition, this rating can be made for time windows considering time-dependent loads. During transportation of a box, it is forwarded to the next node with probabilistic or deterministic decisions based on the rates of a route.

Opposing deadlocks are prevented by the unidirectional system design. Deadlocks because of loops of waiting boxes are not mentioned by Roidl (2012) and Hofmeister et al. (2010). We believe that the frequency of these deadlocks is extremely low in the studied network topologies, which is why ignoring deadlocks is an acceptable strategy.

Dynamic Routing is based on the objective that the box should be transported to the port promising the best route. It can easily lead to deadlocks in networks with long bidirectional lanes, as used for FlexConveyor, because two boxes on opposing routes cannot give way to each other. In dense networks, a dynamic routing is possible because each box can be transported in multiple directions, and thus give way to other boxes. In GridStore, GridPick, GridSequence and the modular warehouse within the Internet Of Things, such dynamic routing algorithms are used.

Depending on how the control algorithms are designed, different deadlock handling strategies are implemented. In GridStore and GridSequence, for example, one conveying direction is excluded system-wide because boxes are never transported to the North. Hence, deadlocks because of boxes waiting for each other in a loop are structurally impossible. The proof of system liveness of GridStore is given with the help of sentential logic and can be found in (Gue et al. 2014). In GridPick, the control algorithm aims to keep a balanced number of boxes per row. Deadlock Handling strategy could be considered to be dynamic avoidance by careful resource allocation. The proof of system liveness of GridPick has been pursued using Petri Net Modeling for a system of up to 5×5 conveyor modules and can be found in (Uludag 2014). Within the modular warehouse described by Sittivijan (2014), conflicts between different boxes are detected and resolved by using priorities. This corresponds to the strategy of detecting deadlocks and recovery with the help of back- or sidetracking of certain boxes. No proof for the absence of deadlocks is presented. Schwab (2015) also detects deadlocks and resolves them by assigning priorities to vehicles. The proof is given with the help of sentential logic. Hama et al. (2002) also use priorities which are dynamically changed according to the history of an agent. They do not go into further details about the handling of deadlocks in loops.

Time-Independent Route Reservation is used for the control of both FlexConveyor and Cognitive Conveyor. The reservation algorithm prevents opposing routes. Nevertheless, deadlocks occurring in loops could still occur.

Within the control of the electric monorail system described by Wilke (2006) these deadlocks are not mentioned by the author. We believe that deadlocks in loops are improbable because of the sparse network and the low number of vehicles in the system. By contrast, complex deadlock handling is performed before every transport step in order to avoid deadlocks occurring in loops if the system design allows the occurrence.

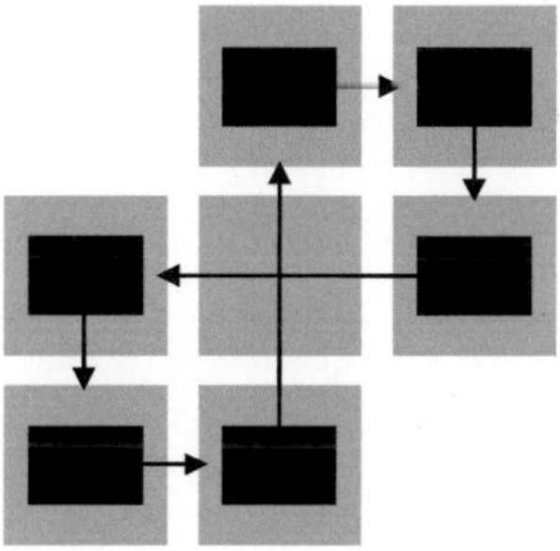

Figure 2.12: Cross-deadlock because of boxes waiting in two overlapping loops

Mayer (2009) handles deadlocks by sending special messages, so-called deadlock tokens, before each transport step. The algorithm is similar to the distributed deadlock detection proposed by Chandy et al. (1983). In order to reduce communication effort, Mayer (2009) uses topology information and only checks for deadlocks if a box enters a loop. This deadlock handling does not avoid so-called cross-deadlocks (see Figure 2.12). Mayer (2009) prevents cross-deadlocks by restricting the routing of boxes: Certain combinations of reservations are not accepted.

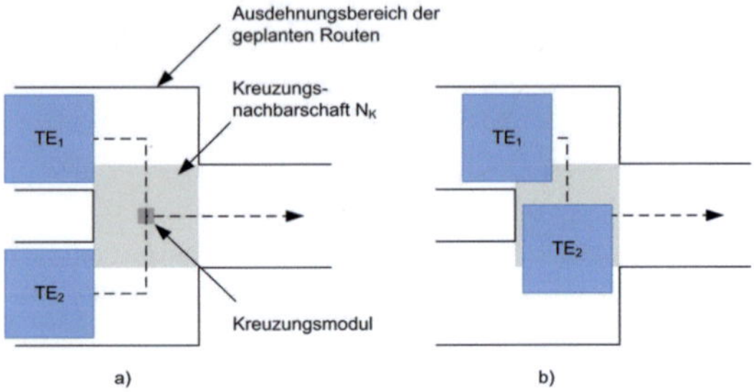

Figure 2.13: Local deadlock because of two boxes blocking each other at a crossing (Krühn 2014)

Krühn (2014) extends the deadlock handling algorithm of Mayer (2009) in two ways: By sending cross-deadlock tokens in certain situation, cross-deadlocks are avoided and the route reservation is not restricted in order to prevent cross-deadlocks. Since the modules of cognitive conveyors are smaller than the transported boxes, Krühn (2014) additionally integrates a mechanism to avoid local deadlocks on crossings (see Figure 2.13). He states that communication effort for deadlock handling is not locally limited and that overlapping crossings must be able to exchange messages with short latencies.

Time-Window-Based Route Reservation structurally negates deadlocks because the route is reserved in advance using time windows. Hence, a conveyor module is only reserved for one box at the same time. Mayer (2009) names the high computing time and high data transfer requirements which would be required for replanning of routes as disadvantages. Time windows would have to be adapted permanently if transports cannot be fulfilled as planned in order to avoid deadlocks. It is not clear whether replanning could always avoid deadlocks.

In material handling systems with decentralized control, it has not been studied how time-window-based routing can be applied. This is why the next section presents an excursus to the research field of time-window-based routing strategies for systems with AGVs.

2.3.3 Excursus: Time-Window-Based Route Reservation in AGV Systems

In section 2.1, we have learned that common characteristics and differences exist between continuous and discontinuous material handling systems.

The routing strategies are similar but they need to be executed differently because the decision is taken by different participators. For an up-to-date overview, we suggest reading the literature review by Vivaldini et al. (2015).

The strategy of time-window-based route reservation has already been applied to systems with AGVs. Routes are computed and reserved for one vehicle after another under consideration of already existing route reservations. The routes are thus computed sequentially. How do the developed routing algorithms deal with the challenge that a route is not executed by the vehicle as previously planned? We will present below the work of four different research groups who take into account the occurrence of incidences such as a delayed vehicle. Briefly said, there are two different methods of dealing with incidences: rerouting of vehicles or reordering of vehicles.

Kim and Tanchoco (1991) present one of the earliest approaches to time-window-based routing and call it *conflict-free routing*. They introduce the notion of a *time window graph* being "a directed graph in which the node set represents the free time windows and the arc set the reachability between the free time windows" (Kim and Tanchoco 1991, p. 2380). The time window graph which is uniquely defined for a specific source node and a starting time is computed during route computation by considering conflicts between different edges. As a path-finding algorithm, they use Dijkstra, but also present an alternative using A*. With regard to incidences, they use so-called *safety allowances* enlarging the time windows reserved for a vehicle. These enlarged time windows enable the system to cope with minor changes in the plan execution but also reduce throughput. If bigger changes in plan execution lie outside of the safety allowance, rerouting becomes necessary for which Kim and Tanchoco (1991) do not present any method.

Gawrilow et al. (2008) present a similar time-window-based routing algorithm which they call *dynamic routing*[2]. In contrast to Kim and Tanchoco (1991), they generate a graph of free time windows on edges and not on nodes. They model the AGVs as polygons and restrict the simultaneous usage of edges if the polygons of two vehicles could collide when generating the time window graph for the whole system. After each route reservation, the time window graph is adjusted accordingly. Similar to Kim and Tanchoco (1991), they use *safety tubes* in order to cope with small incidences like a slightly delayed vehicle. If a delayed vehicle is in conflict with another reservation, they reroute the affected vehicle. Therefore, Stenzel (2008) presents three different rerouting strategies: In the first, the geographical part of the route is kept and only the time windows are adjusted. In the second, an alternative route is found for the part of the route where the incidence occurred. The third one is used if the previous strategies do not succeed: The vehicle is stopped and a new route is calculated from its current position to its destination. Other vehicles could be affected and must also be rerouted. If vehicles block each other from rerouting, they can be routed to so-called parking zones to move out of the way. Like this, successful rerouting is guaranteed.

Maza and Castagna (2005a) present a time-window-based routing which is based on the route planning algorithm of Kim and Tanchoco (1991). They call it *sequence-based conflict-free* routing because of their strategy for dealing with incidences: They use the scheduled entry time of a vehicle in a resource to establish a resource-dependent priority of the vehicle. If this priority is respected i.e. the order of the vehicles using one resource corresponds to the order of the scheduled entry times, deadlocks are prevented. In (Maza and Castagna 2005a; Maza and Castagna 2005b), they propose different methods for changing this priority if incidences occur in order to decrease the impact of incidences.

Context-aware routing is a time-window-based routing algorithm presented by ter Mors (2009) based on the ideas of the three previously presented approaches. For the route planning he uses the *free time window graph* where nodes represent free time windows of resources and edges represent reachability between two time windows (see Figure 2.14). The graph is computed before computation of the route and is not specific for a source node and a start time. The approach of ter Mors (2009) to deal with in-

[2] Unfortunately, they use the term *dynamic* with a different meaning than described in section 2.3.2. To distinguish whether a route is planned before or during execution, they use the terms *offline* and *online* respectively.

cidences is similar to the one of Maza and Castagna (2005a). From the plans of all vehicles, he derives the order in which vehicles will use a specific resource. The timing (or the execution) of plans can be changed without risk of deadlocks as long as this order is respected (ter Mors et al. 2008). He also presents some repair algorithms for changing the order of vehicles. These repair algorithms detect more situations for safe reordering of vehicles than those of Maza and Castagna (2005a). Zutt et al. (2010) present a repair algorithm where the reordering of the vehicles is accomplished by rerouting. They study different heuristics to determine the order in which the vehicles are allowed to reserve a new route. Within the best performing heuristic, the vehicle with the longest wait is allowed to reroute first. They test their rerouting algorithm in grid-based topologies, amongst others, and show that it outperforms the approach without rerouting.

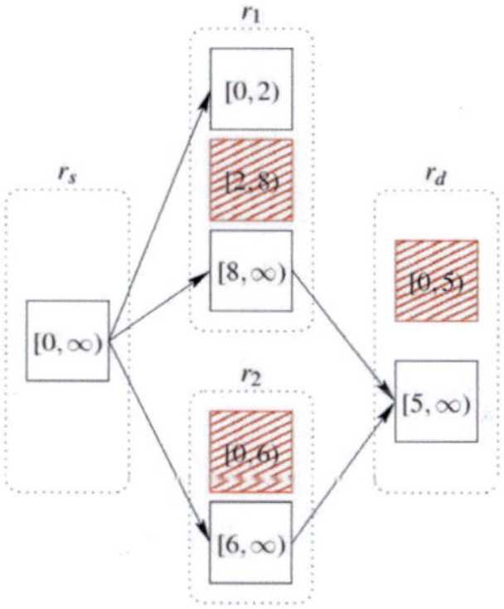

Figure 2.14: Free time window graph where the edges represent reachability of free time windows (Zutt et al. 2010)

What can we conclude from this short overview of existing research in time-window-based route planning for AGV systems? The approaches for dealing with incidences can be divided into two groups. Some of them try to make the planned route more robust by enlarging the reserved time-windows. If conflicts arise because of incidences, a rerouting is triggered. The other approach is to extract the orders of vehicles on resources from the planned route. This order can be either respected in order to prevent deadlocks or the order can be changed with regard to potential deadlocks arising.

What are the differences between AGVs and a modular, continuous material handling system with decentralized control like GridSorter?

- In systems with AGVs, the routes are planned by a central controller with the knowledge of all existing reservations. Even if the route planning is executed by the individual vehicle, the knowledge of all existing reservations is required. Within continuous systems, it is sufficient if each resource only knows its own reservations.
- The routes are planned sequentially for one vehicle after another. Within continuous systems, multiple routes should be planned in parallel.
- If the routes of AGVs are planned by a central control instance, there is no need to synchronize any clocks. The challenge of synchronizing clocks of decentralized controls has not been addressed. The synchronization would become necessary in continuous as well as continuous systems.
- In the presented research for AGVs, all rerouting approaches have been triggered by a central control instance knowing the complete system state (including the current position of each vehicle and the occurrence of incidences). In continuous systems with decentralized control, no control instance knows all reservations or the position of all transported boxes.

2.4 Classification of Material Handling Systems with Decentralized Control

Several classifications exist for some of the systems presented in section 2.2: Schwab (2015) introduces a taxonomy to describe "GridFlow"-systems and Mayer (2009) presents criteria to define the term "completely decentralized". Within this thesis, the principle of logical time is applied to the decentralized control of material handling systems in order to develop an efficient routing and deadlock-free system behavior. This is why we introduce criteria that are related to these challenges in section 2.4.1. An overview of how existing systems can be classified with these criteria is given in section 2.4.2. Parts of this section have been developed together with Geier (2015).

2.4.1 Criteria for Classification

The following classification focuses on the design of the control, especially the routing strategy and the handling of deadlocks in loops. We have already seen that system characteristics such as density of goods or network density influence the risk of deadlocks which is relevant for the choice of deadlock handling strategy. We therefore also include these criteria in our classification.

Main System Task The main system task describes the intra-logistic task the system has been designed for. If a system is suited for multiple tasks, we only included the most important one in the classification. The intra-logistic tasks include *Transport*, *Sorting*, *Storing*, *Buffering*, *Sequencing* and *Picking*.

Routing Task Even if the systems have quite different tasks from an intra-logistics perspective, they can be reduced to two different tasks from a routing perspective. Either items have to be transported to certain destinations (routing task *Transport*) or some items have to be buffered/stored while other items need to be retrieved (routing task *Retrieving from buffer*). In the latter case, one of the main tasks of the routing is to solve conflicts with the buffered items.

Goods Density The density of the goods clearly depends on the routing task. If items need to be buffered/stored, a *high* density of goods should be achieved. If items should be transported to a destination, the density of goods is *low* compared to systems with buffered/stored items. If the objective is to achieve high system throughput, a compromise must be found between a high number of items in the system and increased lead times because of items interfering with each other (cf. Little's law).

The risk of deadlocks increases with increasing density of goods.

Network Criteria Related to the network of the material handling system, two criteria are important for routing and deadlock handling: On the one hand, it is crucial whether the network is *unidirectional* or *bidirectional*. On the other hand, the density of the network is important. The network density can be described with the average node degree (which is the number of neighbor nodes of a node). In the existing systems, all grid-based networks achieve an average node degree close to four neighbors[3]. Theoretically, cognitive conveyors have a higher node degree because they enable omni-directional movement of the boxes, but the developed routing algorithm only includes the four cardinal directions. The network density is *high* if most of the modules are connected to four neighbors and *low* if modules are usually connected to fewer than four neighbors.

Depending on the combination of the two criteria, the risk of different deadlock types is high. The risk of opposing deadlocks is high in bidirectional, sparse networks, whereas the risk of deadlocks in loops increases with the network density independently of unidirectional or bidirectional usage.

Routing Strategy In section 2.3.2, we introduced four different routing strategies: *Forwarding*, *Dynamic Routing*, *Time-independent route reservation*, *Time-window-based route reservation.*

Loop Deadlock Handling In section 2.3.1, we introduced four different deadlock handling strategies: *Ignoring*, *Detection & Recovery*, *Avoidance* and *Prevention.* We limit this criterion to the handling of deadlocks in loops because opposing deadlocks are less complex to handle.

2.4.2 Classification of Existing Systems and Deductions

Table 2.2 shows the classification for all systems presented in section 2.2. The continuous material handling systems are shown in the upper part of the table and the discontinuous systems in the lower part. For better readability, the systems are first divided by the criteria **Routing Task** and **Goods Density** and then ordered by the criterion **Loop Deadlock Handling**. What can be deduced from this overview?

With regard to the **Routing Strategy**, we can state the following:

- *Forwarding* is only used in *unidirectional* networks because opposing deadlocks are prevented by the system design.
- *Dynamic Routing* is only used for *Retrieving from buffer.* It seems to be easier to solve conflicts with buffered items by dynamic negotiation with the local neighbors.

[3] The average node degree in grid-like patterns is not equal to four because border modules have less than four neighbors

System	Main System Task	Routing Task	Goods Density	Network Criteria	Routing Strategy	Loop Deadlock Handling
Internet of Things	Transport	Transport	low	sparse & unidirectional	Forwarding	Ignore
EHB	Transport			sparse & bidirectional	Time-independent route reservation	Ignore
FlexConveyor	Transport			sparse & bidirectional	Time-independent route reservation	Avoidance
Cognitive Conveyor	Transport			dense & bidirectional	Time-independent route reservation	Avoidance
GridSorter as of Seibold et al. (2013)	Sorting			dense & bidirectional	Time-independent route reservation	Incomplete Avoidance
Modular Warehouse	Buffering	Retrieving from buffer	high	dense & bidirectional	Dynamic routing	Detection & recovery
GridFlow with AGVs	Buffering					Detection & recovery
Cellular Warehouse	Buffering					Detection & recovery
GridPick	Picking					Avoidance
GridStore	Storing					Prevention
GridSequence	Sequencing					Prevention

Table 2.2: Classification of material handling systems (MHS) with decentralized control

- *Time-independent route reservation* is used for *Transport* of items in *bidirectional* networks.
- *Time-window-based route reservation* is not used (yet).

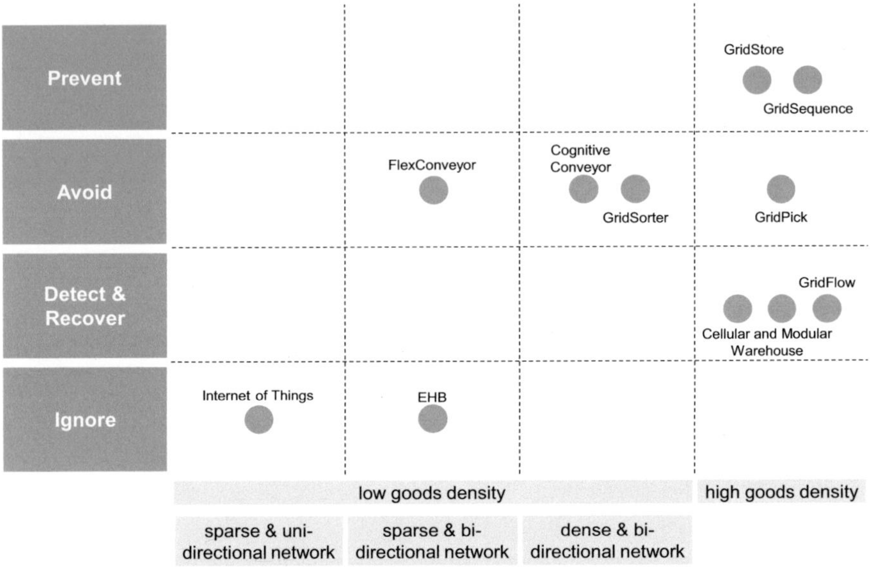

Table 2.3: Deadlock handling strategy in relation to goods density and network criteria

Table 2.3 illustrates the relation between goods density and network criteria on one side and the deadlock handling strategy for loop deadlocks on the other side. We can state the following:

- *Ignoring* deadlocks is only applied in systems where network density and goods density are *low*.
- *Detection & Recovery* is only applied in *dense*, *bidirectional* networks because recovery is feasible with sidetracking of an item in any direction. Moreover, it is only applied for *Retrieving from buffer*, because a buffered item might be used for sidetracking with no or low impact on the system performance.
- *Avoidance* seems to be generally suited to different systems, but one of the presented algorithms does not avoid all deadlock situations.
- No *Prevention* algorithm exists for *Transport* of items.

We can see that the choice of deadlock handling strategy is closely related to the network criteria and the goods density. Table 2.3 illustrates: The higher the network and goods density, the higher the risk of deadlocks, the "higher" the chosen deadlock handling strategy.

2.5 The GridSorter as Showcase System

We have chosen the GridSorter as showcase system because it is suited to the implementation of a control algorithm for the transport of goods to specific destinations in dense networks with high deadlock risk.

Figure 2.15 shows the physical system: Each of the modules is a conventional 90°-transfer module and is able to convey a box in four directions. With optical sensors on all four sides of the module, it detects the position of the transported good and can thus control the transport of the goods. Using its control, called FlexBox, the module can independently process information and make decisions. A physical, electrical and electronic connection exists to all four directly adjacent modules. Two adjacent modules can therefore exchange information in form of messages.

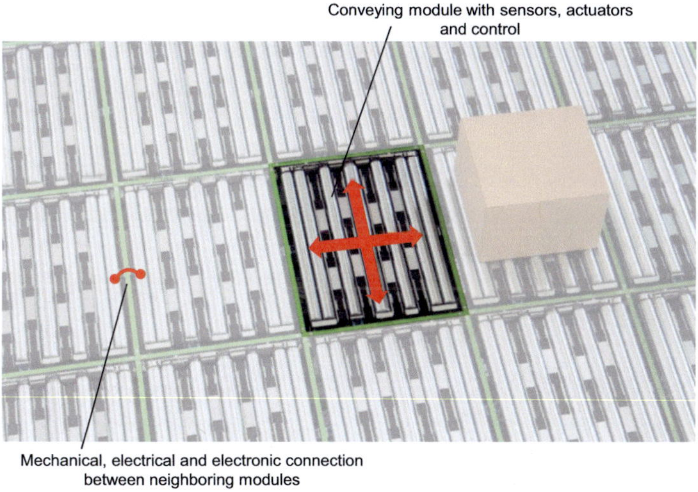

Figure 2.15: The physical system: a conveying module and its connections

The topology of the network is recognized by a decentralized process based on the Link-State-Routing-Algorithm (Tanenbaum 2011): After each module has requested the identifier of its neighbors, this information is distributed through the entire network. With this information, each module generates an adjacency matrix of the network. In this way, it can calculate shortest paths and extract the minimal path length to destinations, for example. Information about the goods such as the destination can be achieved in different ways: The relevant information can either be read from a bar-code/RFID chip or it is provided by an external system. Once the source module knows the goal of the carried box, it starts, depending on the control algorithm, actions for route reservation or transport of the box.

2.6 Conclusion: The Research Gap

There are various material handling systems with decentralized control. Some of them are integrated in visions like the Internet of Things or the Physical Internet. The field of systems with decentralized control seems to be very active because a lot of the presented systems have been newly developed in the last five years (compare to systems presented by Mayer (2009)). Two systems are even industrialized (FlexConveyor and THINGtelligence). All of the presented systems correspond well to the design rules formulated in the vision of the Fourth Industrial Revolution (Kagermann et al. 2012).

The system design, especially the density of the network and the transported goods, influences the risk of deadlocks. A lot of effort has been invested in developing control algorithms that handle deadlocks in dense networks. Nevertheless, no deadlock prevention algorithm exists for systems where all items have to be transported to a specific destination as fast as possible. The routing strategy is closely related to the deadlock handling strategy. To date, no decentralized control algorithm exists in the presented material handling systems that uses time windows for the route reservation.

The objective of this thesis is to close the research gap by developing a control algorithm that prevents deadlocks and can be applied to a wide range of system tasks. The GridSorter is used as showcase system to apply the developed control algorithm.

3 Distributed Deadlock Prevention with Logical Time

Pure mathematics is, in its way, the poetry of logical ideas.
— Albert Einstein

In the previous chapter, we presented how existing decentralized control algorithms of material handling systems deal with the risk of deadlocks. In section 3.1, we reason our choice of routing strategy. After presenting a formal system description in section 3.2, we then introduce the principle of logical time (section 3.3) and present the control of GridSorter using logical time (see section 3.4). In section 3.5, we show that the clock condition formulated by Lamport (1978) is satisfied following the reservation and granting conditions. Finally, in section 3.6, we prove the absence of deadlocks if this principle is used. Parts of this chapter have been published by Seibold and Furmans (2014).

3.1 Choice of Routing Strategy

Section 2.3.2 presented existing material handling systems with decentralized control. Deadlock risk increases with the network density and the density of transported goods. As the modules of GridSorter form a dense network, the risk of deadlocks is present. Consequently, the strategy of ignoring deadlocks is not an option.

We examine if and how the three presented routing strategies could be applied to GridSorter and how the mechanisms for deadlock handling impact the system performance and complexity of the routing control.

Forwarding The routing strategy to forward boxes in the most promising direction has only been applied in unidirectional networks because it does not handle opposing deadlocks in bidirectional networks. Within GridSorter, additional control rules would be required to handle opposing deadlocks and deadlocks in loops leading to complex deadlock avoiding mechanisms.

Dynamic routing Within dense networks, dynamic routing can be applied because each module has enough routing possibilities to give way for other boxes. Since the routing decisions are taken short-term, only the current state of the local surrounding is taken into account. Therefore, it is hard to guarantee efficient system behavior because future system states are not considered. Also, the risk of livelocks and starvation increases. Dynamic routing has only been applied to systems where a larger proportion of the boxes needs to be stored. Within GridSorter, the objective is to transport all boxes to their destination. Hence, it is possible to take future system states into account. System liveness is hard to prove in systems with dynamic routing decisions because interdependency of multiple local decisions must be taken into account.

Time-Independent Route Reservation With the strategy of time-independent route reservation, deadlocks must be avoided during the transport of a box by sending messages to detect loops of waiting boxes. In a dense network like GridSorter, the number of needed message transfers before each transport step increases because the loops can have any form and size in a dense network. In addition, the communication for deadlock avoidance increases with system size because it is not locally restricted. In large systems, transport delays could occur because of a high number of needed message transfers. On the other hand, the deadlock avoidance algorithm is very complex, as can be seen in Krühn (2014), which increases the effort for a correct implementation and adaption to different systems and system tasks.

Time-Window-Based Route Reservation With time-window-based route reservation, waiting times can be taken into account during the routing process. Using time windows of physical time in a system with decentralized control, the physical clocks of the modules have to be synchronized, which could be done by communication. Also, it is a challenge to handle transport delays leading to arrival times at modules differing from the planned time windows. One approach in AGV systems is to increase the reserved time windows to reduce the probability that a reserved time window has to be adapted (Stenzel 2008). Another approach is to ignore the reserved time windows and only to respect the order of vehicles using a specific resource for preventing a deadlock (ter Mors 2009). Within GridSorter, this would mean that an order must be established between all boxes transported by the same conveyor.

Based on this assessment, we have chosen the time-window-based route reservation as routing strategy. What kind of principle could be used in order to combine the advantages of time-window-based route reservation (efficient and deadlock-free routing) without incurring the disadvantages (synchronization, rerouting)? How can the ordering of transports be established and respected in a system with decentralized control?

Partial ordering of resources has been studied in the field of distributed multi-process-systems by Lamport (1978). He uses logical clocks to establish a partial ordering of events in order to represent causal relations between events. In the next section, we therefore describe our material handling system as a multi-process system with multiple resources.

3.2 Formal System Description and Assumptions

The system consisting of conveyor modules and boxes can be considered to be a distributed multi-process system as follows: The transport of a box from source to destination is considered to be a process. When the box is being conveyed i.e. moving, the corresponding process is in state *Transport*, otherwise it is in state *Hold and Wait*. The process starts when the box is inserted in the system and the transport steps from one conveyor module to the next one are considered to be events of this process. The process is finished when the box reaches its destination.

Conveyor modules are considered to be resources. A box needs to be conveyed by two conveyor modules concurrently to be transported from one module to the next one. Thus, the corresponding process in state *Transport* must hold to both resources. The duration of this transport phase is not defined but finite. When the box is standing still, it remains on one conveyor module. Thus, a process in state *Hold and Wait* only needs to hold to one resource but it has already requested the resource it needs for the next transport step. Usually, a resource is only assigned

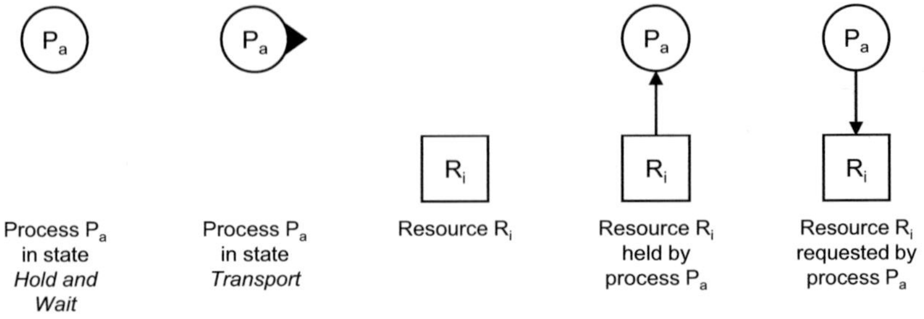

Figure 3.1: Legend for figures explaining resource acquisition

to one process, because a conveyor module is the same size as one box. Tandem transport is the only exception: Having three aligned conveyor modules, the middle module can receive a box from one side and send a box to the opposite side concurrently. In the case of tandem[1] transport, this resource is assigned to two processes.

One important difference between the system of modular conveyors and a distributed multi-process system lies in the characteristic of Cyber-*Physical*-Systems: The time required to *physically* move a box is many times higher than the time needed to share and process information between distributed controls in *cyber*-space. This is also the reason why the route can be reserved for the transport of each individual box.

The graphical representation of processes, resources and possible relations between them is shown in Figure 3.1 and is based on the representation of Tanenbaum and Bos (2015).

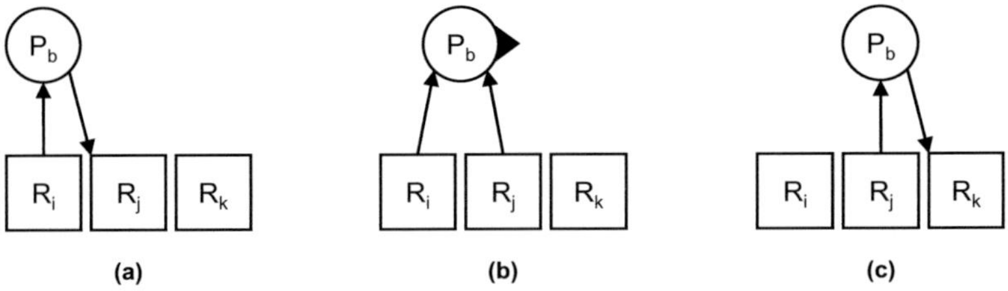

Figure 3.2: Course of acquisition of resources for single transport

[1] The literal meaning of the word "tandem" implies that only two boxes move together. Nevertheless, if several aligned modules perform tandem transport simultaneously, a chain of boxes (more than two) move together. This is also called slug movement.

Tanenbaum and Bos (2015) describe a sequence of actions required to use a resource as follows: request, use and release the resource. Because of the route reservation, the sequence of actions is the following in our system:

1. Reserve the resource
2. Request the resource
3. Use the resource
4. Release the resource

What is the sequence of actions for the transport of a box? Figure 3.2 shows the course of acquisition of resources for single transport. All resources have been reserved already. Part (a) shows process P_b holding to resource R_i and having requested resource R_j. R_j not being assigned to any other process, it is granted to process P_b which then switches to state *Transport* holding both resources (see part (b) of Figure 3.2). After a finite duration of transport, process P_b releases resource R_i and requests the next resource R_k which is shown in part (c). We note that a resource can only be released by a process if the next resource has been granted to this process already.

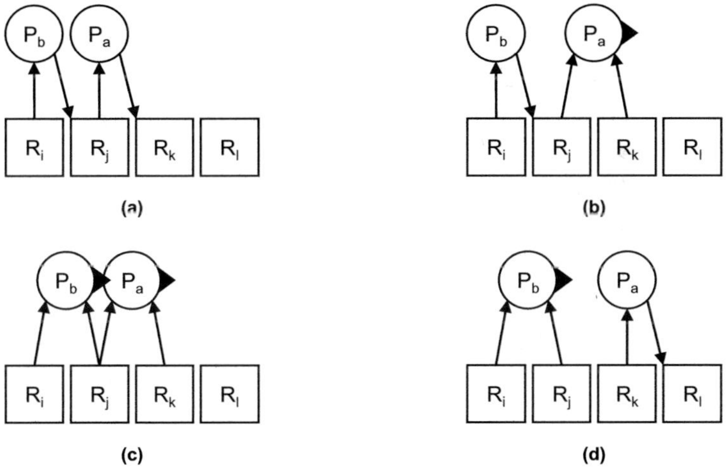

Figure 3.3: Course of acquisition of resources for tandem transport

Figure 3.3 shows the course of acquisition of resources for tandem transport. Again, the resources have been reserved already. In part (a), process

P_b is again requesting resource R_j which is this time already held by process P_a. Process P_a, in turn, has requested resource R_k. Tandem transport is physically possible because the trio of resources R_i, R_j and R_k is aligned. Once resource R_k is granted to P_a, P_a switches to state *Transport* (see part (b)). Resource R_j can now be granted to process P_b while it is still held by P_a (see part (c)). P_b switches to state *Transport* as well. Transition to part (d) is triggered when process P_a switches to *Hold and Wait* after a finite physical time interval. In addition, process P_b switches to *Hold and Wait* and requests the next resource after a finite physical time interval. This situation is not shown in the figure.

Let us summarize: The transition from state *Hold and Wait* to *Transport* takes place when granting conditions are fulfilled. A resource can only be granted to a process if it is free, i.e. not held by another process, or if tandem transport conditions are fulfilled. The transition from *Transport* to *Hold and Wait* is time-triggered because the transport has a finite duration in physical time.

The property of system liveness includes the termination of each process i.e. each box must be transported to its destination. In order to prove system liveness, it is necessary to prove the absence of deadlocks, livelocks and starvation.

3.3 Transferring Logical Time to Material Handling

A distributed system consists of multiple parallel processes, each being a set of events with a defined order. Because of this defined order, causal relations exist between the events of one process. Lamport (1978) calls a causal relation between two events a "happening before" relation. The events of the processes are represented by dots in Figure 3.4. The parallel processes are connected to each other by messages defining a causal relation between the events of two different processes: The event of sending a message happens before the event of its reception. The sending of a message is represented by a wavy arrow in Figure 3.4. The entire set of causal relations forms a partial ordering of the events in the system. Event p_1 is happening before q_3 in Figure 3.4 because they are connected through a sequence of causal relations. However, the ordering of event p_3 and q_3 cannot be determined because there is no sequence of causal relations.

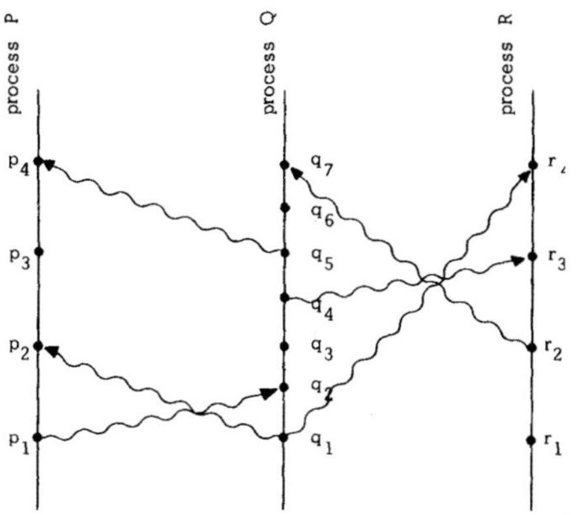

Figure 3.4: Distributed system: parallel processes with multiple events and causal relations (Lamport 1978)

Lamport (1978) assigns a number to each event representing the logical time at which the event occurs. C_i is the logical clock of process P_i and $C_i\langle a\rangle$ the logical time of event a if it is an event of process P_i. The set of logical clocks of all processes is represented by C where $C\langle a\rangle = C_i\langle a\rangle$ if a is an event of process P_i. He formulates the clock condition as follows:

For any events a, b: if $a \to b$ then $C\langle a\rangle < C\langle b\rangle$

where "$\to$" is the symbol for "happening before". Please note related to the clock condition that the logical time of an event is assigned by the set of all logical clocks C.

Let us now transfer the principle of logical time to the decentralized control of a material handling system. First, we present the parallelisms between a modular conveyor system and distributed systems.

- Parallel processes exist because multiple boxes should be transported concurrently.
- The transport steps needed to convey the box to its destination represent the events[2] of this process.
- There are "happening before" relations among the events of one process, i.e. the transport steps of one box, arising from the physical movement.

[2] Usually, an event is not allowed to have a duration. In our case, the duration is defined in physical time. In logical time, the event does not have any duration.

- For each event of one process, two resources are needed, because one transport step is enabled by two conveyors.
- If routes of two different boxes overlap each other, the conveyors used by both boxes are common resources of these processes[3].
- There are "happening before" relations among the events related to one resource used by multiple processes, i.e. the transport of different boxes on the same conveyor.
- There are no "happening before" relations between boxes whose routes do not overlap. The transport steps of all boxes can thus be partially ordered.

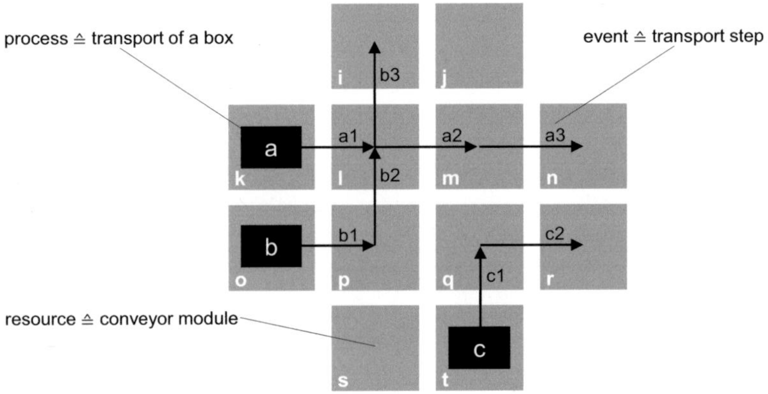

Figure 3.5: Example for routes of three boxes with one common resource

Figure 3.5 shows an exemplary situation of three boxes on the GridSorter. As described before, two different kinds of "happening before" relations exist:

First, the transport steps of each box can only take place in a predetermined order.

$$a1 \rightarrow a2 \rightarrow a3$$
$$b1 \rightarrow b2 \rightarrow b3$$
$$c1 \rightarrow c2$$

[3] If tandem transport is possible, a common resource is held by two processes concurrently. If single transport is required, a common resource must be used by the processes subsequently.

Second, on the module where the routes of box a and b cross, the transport steps must happen in a certain order: Box a must have left the module before box b enters or vice versa.

$$b3 \rightarrow a1 \text{ or } a2 \rightarrow b2$$

The route of box c does not overlap with another route. Therefore, no causal relations exist and the transport of box c can be performed independently from the other boxes.

For the two kinds of "happening before" relations on GridSorter, the following two conditions must be fulfilled in order to satisfy the clock condition:

1. The transport steps of one box must happen at advancing logical time.
2. Consider one conveyor module: The outgoing transport of one box must take place at an earlier logical time than the incoming transport of the next box.

In the example of Figure 3.5, the following conditions must be fulfilled regarding the logical time at which the transport steps are performed:

$$C\langle a1\rangle < C\langle a2\rangle < C\langle a3\rangle$$
$$C\langle b1\rangle < C\langle b2\rangle < C\langle b3\rangle$$
$$C\langle c1\rangle < C\langle c2\rangle$$
$$C\langle b3\rangle < C\langle a1\rangle \text{ or } C\langle a2\rangle < C\langle b2\rangle$$

One possible combination of logical times of the events satisfying these conditions is shown in Figure 3.6.

We want to use the principle of the logical time for partial ordering of transport steps to design a deadlock prevention that does not have an negative impact on the system performance. In the next section, we describe how material handling systems could be controlled using logical time.

3.4 Controlling Material Handling with Logical Time

Lamport (1978) describes an algorithm using logical time for a system where parallel processes send messages to each other containing timestamps T in order to request access to a common resource. The logical clocks are assigned to the processes and set according to the timestamps contained in received messages.

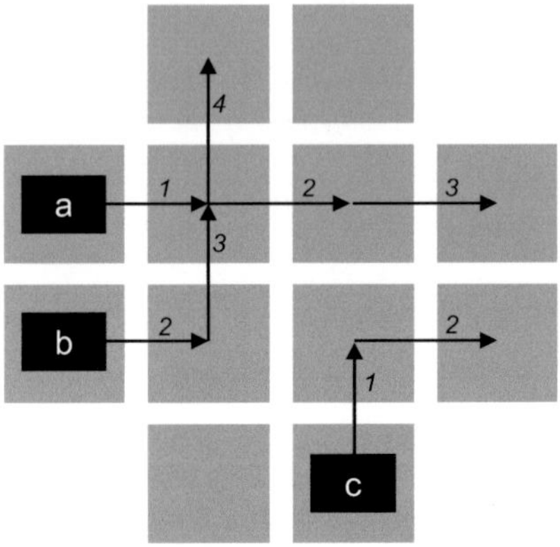

Figure 3.6: Example for routes of three boxes with exemplary logical times

In our system, multiple common resources exist and each process can be terminated by using a different sequence of resources. In addition, it is not the processes that send messages and take decisions, but the resources[4]. This is why the control algorithm presented by Lamport (1978) needs to be extended in order to be applied to the control of material handling. In section 3.4.1, we present our solution of assigning logical clocks to resources instead of processes.

Figure 3.7 shows for each box, resources are reserved using timestamps based on the described "happening before" relations (section 3.4.2). During the transport process, the partial order among the transport steps defined by the set of reservations has to be respected. Therefore, we describe the conditions for requesting, granting and releasing of resources in section 3.4.3. These conditions are closely related to the timestamps of the corresponding reservations and guarantee that the logical clocks of the resources are only set forward and never back. Figure 3.7 also shows

[4] This is the point where I want to say special thanks to my sister and her husband for their patience during hour-long discussions whether the boxes should be the processes or the resources. I learned that for IT specialists it is very contra-intuitive that a process is a dumb thing, incapable of taking decisions or having a logical clock. However, we model our system like that because all other parallelisms fit very well.

that the boxes enter the system at different times. Thus, the processes are not synchronized and a resource must be able to receive a new reservation while it is held by another process and while events are taking place. Basic ideas of this section have been developed together with Haude (2014).

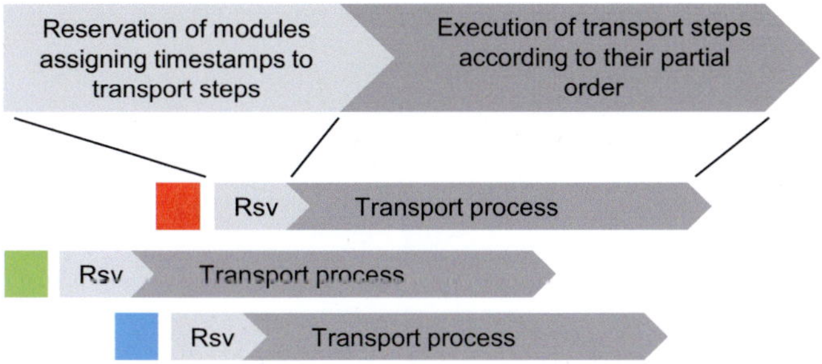

Figure 3.7: Parallel reservation of resources and execution of transport steps

3.4.1 Assigning Logical Clocks to Resources

Figure 3.8 shows the example of Figure 3.5 as a multi-process system inspired by the representation of Lamport (1978). The three processes P_a, P_b and P_c are represented by vertical lines. Thus, each line represents the transport of one box and therefore needs an uninterrupted sequence of resources to be successfully terminated because a box always needs to be carried by at least one conveyor. During a transport step, represented by a dot in the figure, the box is transported from one module to another. For this transport both resources are needed. In the model, the physical duration of a transport step is irrelevant: the event simply includes the process switching from one resource to the next. A “happening before” relation between two processes is represented by a dotted arrow and becomes necessary if a resource is used in common i.e. if two routes cross. In the example of 3.5, there is only one conveyor module which is used by two boxes. In Figure 3.8, the common resource R_l is highlighted in gray. The dotted arrow states that R_l must first be released by process P_a before being granted to P_b which corresponds to the “happening before” relation $a2 \rightarrow b2$.

Even though Figure 3.8 looks very similar to 3.4, it contains an additional dimension compared to the system described by Lamport (1978). For this purpose, please consider resource R_l highlighted in gray in Figure 3.8: Following this resource in the horizontal direction, we recognize that it also forms a sequence of events like a process. Consequently, each event is related to one process and two resources that need to be synchronized.

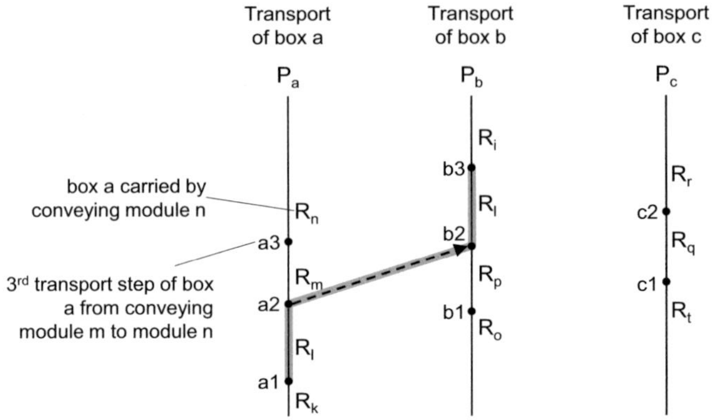

Figure 3.8: Representation of GridSorter as multi-process-system inspired by Lamport (1978)

How can the logical time be assigned to the events in our system? C is the set of all logical clocks in the system and $C\langle a1\rangle$ the number assigned to event $a1$ by the set of all logical clocks. In (Lamport 1978), $C\langle a1\rangle = C_a\langle a1\rangle$ is assigned by the logical clock of process a if event $a1$ is part of process a. In our system, it is not the processes that are equipped with a control, but the resources. Let C_i denote the logical clock of resource R_i. Consequently, the logical time of an event can only be assigned by the related resources which must have logical clocks of their own[5]. For each event of a process, two resources are necessary.

For example in Figure 3.8, the transport step $a1$ of box a is enabled by conveyors k and l. The logical clocks of both resources must assign the same logical time to the event. It must hold good that

$$C\langle a1\rangle = C_k\langle a1\rangle = C_l\langle a1\rangle$$

if $a1$ is an event for which resource R_k and R_l are needed.

[5] Implicitly, the process also has a logical clock having the same logical time as the resource(s) it is holding to.

One might have the impression that the idea of logical time has been drastically changed in comparison to Lamport (1978), but we only expanded the set of all logical clocks C and changed the way in which this set assigns a logical time to an event.

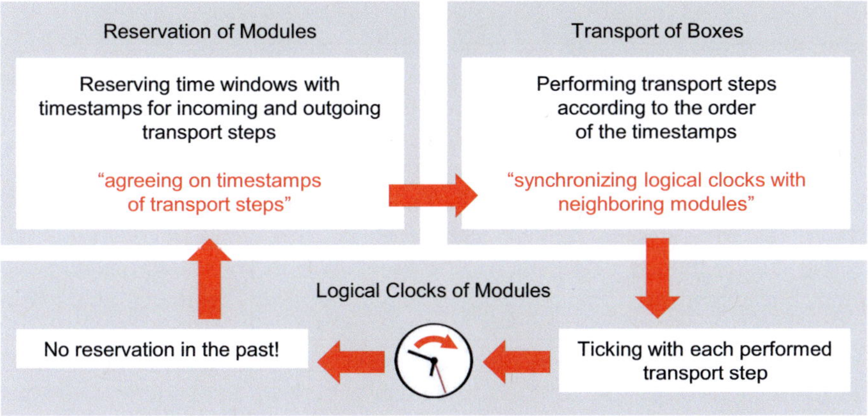

Figure 3.9: Control design of GridSorter with logical time

So how is a valid sequence of resources for a process found? And how do the resources agree on the logical time of an event? The interrelation between the reservation of modules, the transport of boxes and the logical clocks is shown in Figure 3.9. In order to find a route for a specific box to its destination, the conveyors send reservation requests to each other, assigning timestamps T to the transport steps of the box. By assigning a timestamp to each transport step, both conveyors agree on the logical time at which the transport should be performed. The entire set of reservations establishes a partial ordering among all transport steps of all boxes. In order to satisfy the clock condition formulated by Lamport (1978), the transport steps must be performed according to the partial ordering defined by the reserved timestamps. During transport of the boxes, each module sets its logical clock to the timestamp of the performed transport step, and therefore synchronizes its logical clock with the neighboring conveyor module. The transport of boxes influences the reservation of modules because new reservations must only be accepted if they lie in future logical time in order to satisfy the clock condition (see section 3.5).

3.4.2 Reservation of Resources using Logical Timestamps

For each process, all resources needed for its successful termination must be reserved before the process requests the first resource that could be commonly used. To use the principle of logical time for the control of a decentralized material handling system, the reservations must fulfill certain conditions which are introduced in this section. We do not present the algorithm showing how a set of possible reservations is found (i.e. the route reservation process) in this chapter because it is not relevant for the principle of logical time. A possible implementation of the route reservation process is introduced in Chapter 4. In this chapter, we simply assume that a valid sequence of resources is known and reserved for each process. Each reservation is defined by a starting and an ending timestamp because the module is occupied by the box between the incoming and the outgoing transport. Two subsequent conveyor modules on the route of a box agree on the timestamp for this transport step. These reservations are distributively stored among the resources so that each resource only keeps its own reservations. A resource is able to keep multiple reservations for different processes.

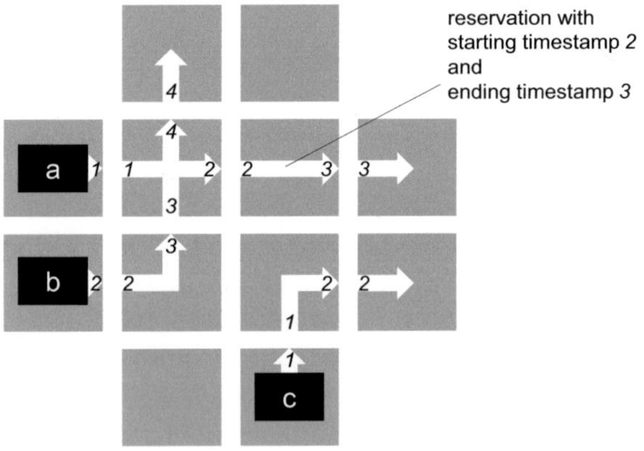

Figure 3.10: Example for routes of three boxes with reservation of resources

Figure 3.10 shows the reservations for the previous example if the modules agree on the timestamps as shown in Figure 3.6. The reservations are represented by white arrows. Each timestamp of a transport step is listed as ending timestamp of the reservation of the sending module and as starting timestamp of the reservation of the receiving module. On the module where two routes are crossing, the reservations do not interfere with each other because the reservation for box a has the ending timestamp *2* whereas the reservation for box b has the starting timestamp *3*.

Please note that the reservations are not necessarily received and accepted in the order of their timestamps. It is possible that a reservation is accepted even though another reservation exists for later timestamps as long as it lies in the future and does not interfere with other reservations. For example, the commonly used resource in Figure 3.10 could have received the reservation of box b before the reservation for box a.

Let us now describe the reservation conditions formally: The reservation of a resource for a specific process is defined by two timestamps: Let $T_{in}(P_a, R_i)$ denote the timestamp of the incoming transport and $T_{out}(P_a, R_i)$ the timestamp of the outgoing transport. The reservation of resource R_i for process P_a is defined by the pair of timestamps

$$[T_{in}(P_a, R_i), T_{out}(P_a, R_i)]$$

with

$$T_{in}(P_a, R_i) < T_{out}(P_a, R_i) \tag{3.1}$$

representing that the incoming transport of a box must take place before its outgoing transport (see Figure 3.11). This condition guarantees that the clock condition for the causal relation between the events of one process is respected.

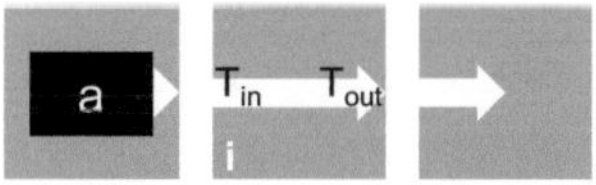

Figure 3.11: Timestamps of one reservation

Two conveyors are needed to transport a box from one conveyor to the next one. If process P_a needs to use resource R_i and R_j subsequently, the condition

$$T_{out}(P_a, R_i) = T_{in}(P_a, R_j) \tag{3.2}$$

must be fulfilled (see Figure 3.12). This condition guarantees that both resources agree on the logical time of the related event.

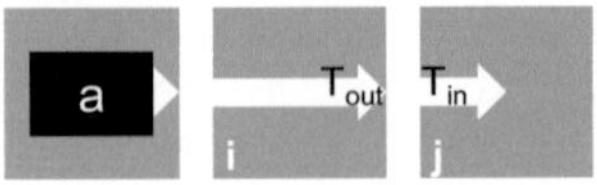

Figure 3.12: Timestamps for one transport step

As stated above, a resource can only be granted to one process at a time (with the exception of tandem movement). If process P_a should use resource R_i before process P_b, the reservations must fulfill the condition

$$T_{out}\left(P_a, R_i\right) < T_{in}\left(P_b, R_i\right) \tag{3.3}$$

if single movement is planned, because box a must have left conveyor i before box b can enter (see Figure 3.13). This being the case, the clock condition for the causal relation between the events of two different processes is respected. Single movement is necessary if the incoming direction of box b is perpendicular to the outgoing direction of box a.

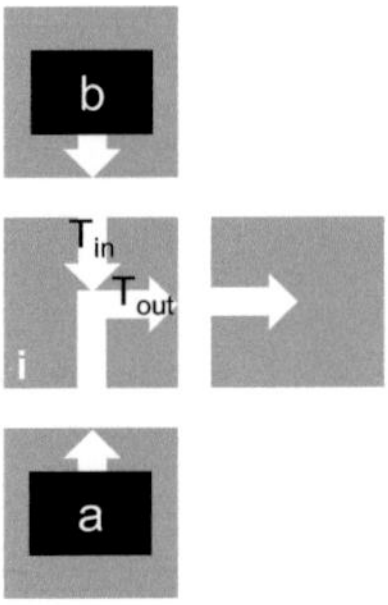

Figure 3.13: Two boxes using module i in single movement[6]

Tandem movement is physically possible if the corresponding trio of resources is aligned, which is why box a leaves module i on the opposite side of box b entering (see Figure 3.14). In this case, resource R_i can be assigned to two processes and instead of condition (3.3), the condition

$$T_{out}\left(P_a, R_i\right) = T_{in}\left(P_b, R_i\right) \tag{3.4}$$

must be fulfilled.

[6] The incoming direction of box a is irrelevant for condition (3.3) and (3.4). The figure only shows an exemplary incoming direction of box a.

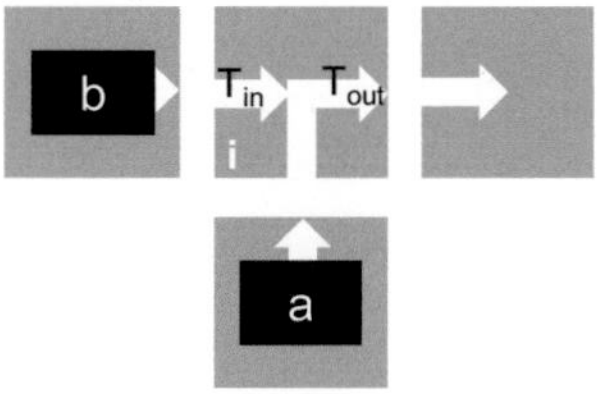

Figure 3.14: Two boxes using module i in tandem movement[6]

Let C_i be the logical clock of resource R_i defining its current logical time. Related to the logical clock, the reservation for process P_a must satisfy the condition

$$C_i \leq T_{in}\left(P_a, R_i\right) \tag{3.5}$$

in order to be in future logical time of the resource[7].

3.4.3 Acquisition of Resources using Logical Time

In section 3.2 we describe the course of acquisition of resources and state that a resource is only granted if the granting conditions are fulfilled. In this section, the conditions for granting and releasing a resource are defined using the timestamps of the related reservation. Each resource holds a list of all accepted reservations. Reservations are deleted once the corresponding transport steps are fulfilled.

Figure 3.15 shows the course of acquisition of resources for a single transport like Figure 3.2 with the additional information of the logical clocks of the related resources. In part (a), process P_b is holding to resource R_i and its logical clock depicts the timestamp of the outgoing transport. If now resource R_j is granted to P_b, it sets its logical clock to the timestamp of the incoming transport.

$$C_j := T_{in}\left(P_b, R_j\right) \tag{3.6}$$

In part (b) of Figure 3.15, P_b holds to resource R_i and R_j and switches to state *Transport*. The logical clocks of both resources show the same logical

[7] As in any calendar system, reservations lying in the past cannot be accepted. An alternative, later time has to be found for the reservation.

time. When the box is successfully transported to the next resource as shown in part (c) of Figure 3.15, the following actions take place:

- The resource R_i is released.
- R_i deletes the reservation of process P_b because all related transport steps have been performed.
- Process P_b requests the next resource R_k and switches to state *Hold and Wait.*
- R_j sets its logical clock to the timestamp of the outgoing transport.

$$C_j := T_{out}(P_b, R_j) \tag{3.7}$$

Please note that the transition from the state *Transport* to the state *Hold and Wait* is only triggered by the end of the transport step i.e. it takes place after a finite physical time without any special conditions related to the reservation.

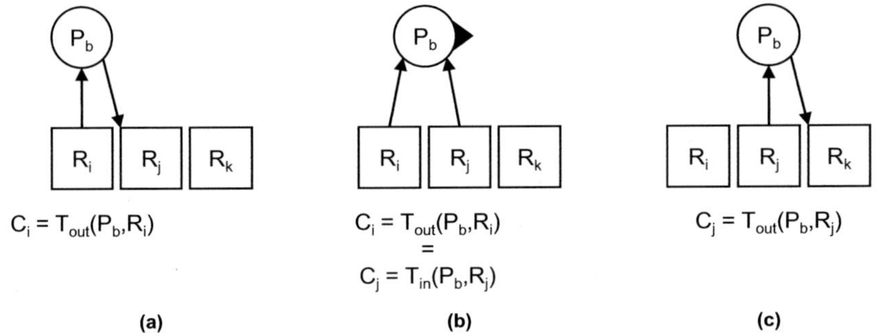

Figure 3.15: Course of acquisition of resources for single transport with logical clocks

What are the conditions for resource R_j to be granted to process P_b? The following three granting conditions (GC) must be fulfilled before starting a transport:

GC1 R_j is requested by P_b. This condition guarantees that the sending conveyor is ready for the transport to be started.

GC2 R_j has performed all transport steps with lower timestamps and deleted the relevant reservations. Therefore, R_j can be granted to P_b if it does not keep a precedent reservation for any process P_a with $T_{out}(P_a, R_j) < T_{in}(P_b, R_j)$[8]. Like this, the conveyor only performs the transport step with the lowest timestamp. This guarantees that the transport steps of the accepted reservations are performed in the previously defined order.

GC3 The third granting condition depends on the transport type:

single R_j is free, i.e. not held by any other process.

tandem R_j is held by P_a which is in state *Transport* and tandem transport is physically possible (compare Figure 3.3).

This condition is related to the physical feasibility of a transport and guarantees that no collisions occur.

The granting conditions guarantee that the transports are fulfilled in the order of their reservation timestamps so that the logical clock of the conveyor is always set forward and never back.

What does this set of logical clocks look like for an external observer? Each module has its own logical clock which is only set forward in discrete steps if a box is transported. If two neighboring modules perform a transport together, their logical clocks are synchronized i.e. their logical clocks are set forward to the timestamp reserved for the transport step. The set of logical clocks does not define *one* identical system time, because all logical clocks could differ from each other. The logical time is completely independent of physical time. For example, it could happen that a conveyor module remains in the same logical time for a long period of physical time because no box has to be transported.

Figure 3.16 shows an conveyor network forming a crossing where boxes have only been transported from WEST to EAST several times. All other modules have remained in the logical time *0*. The left side of the figure shows a valid reservation because all timestamps lie in the future logical time of the participating conveyor modules. When the transport of box a is finished, the conveyor modules have updated their logical time by synchronizing with the neighboring module when performing a transport step. The logical clocks have skipped all time steps in between. To put it more explicitly, one could say that the boxes bring the time to the conveyors.

[8] Within tandem movement, resource R_j keeps a reservation with $T_{out}(P_a, R_j) = T_{in}(P_b, R_j)$ which is allowed.

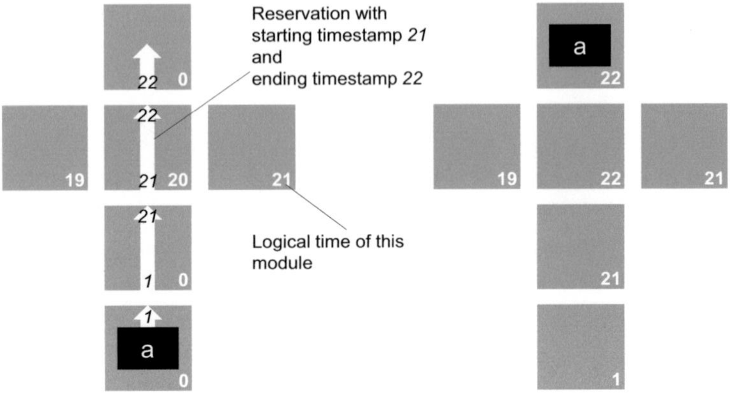

Figure 3.16: A conveying network before (left) and after (right) the transport process of box a

3.5 Proof of Satisfying the Clock Condition

In this section, we show that the described algorithm satisfies the clock condition defined by Lamport (1978). The "happening before" relation only exists between events of one process or between events on one resource. It is therefore sufficient to show that the events of one process take place in ascending order of their timestamps and that the logical clock of one resource is advancing i.e. the related events take place in ascending order of their timestamps.

Proof. Let R be the set of all resources in the system and P be the set of all processes in the system. With the reservation conditions (3.1) and (3.2), we guarantee that the timestamps of reservations of one process are ascending. It is physically given that the transport steps are executed in this order. Thus, the events of one process are forcibly executed with ascending timestamps. In addition to these physical characteristics, the events happen in order of their timestamps because the process only requests the next resource when it releases the previous one (compare section 3.4.3). Let us look at P_b is holding to resource R_j: Since all preceding transport steps of process P_b must have been executed, the associated reservations have been deleted. Thus, if process P_b is holding to resource R_j and waiting for the next resource, no reservation exists for process P_b on any other resource $R_i \in R$ with

$$T_{out}(P_b, R_i) < C_j \,. \tag{3.8}$$

Consider now the events related to one resource: The logical clock is set to a new value if the resource is granted to a new process (equation (3.6)) or if the next resource is requested (equation (3.7)). The clock condition is fulfilled if the logical clock of each resource is only set forward. Let us look at resource R_j that has been previously released by process P_a and is requested by P_b. Before R_j is granted to process P_b, its logical clock corresponds to the ending reservation timestamp of process P_a, thus

$$C_j = T_{out}(P_a, R_j) \,.$$

For the logical time to be advancing, the condition

$$C_j \leq T_{in}(P_b, R_j)$$

must be satisfied if resource R_j is granted to process P_b.

Let us assume the contrary: The logical clock is set backward because $C_j > T_{in}(P_b, R_j)$. Either reservation of process P_b has already existed when the resource has been granted to process P_a. In this case, it would have violated granting condition **GC2** with

$$T_{in}(P_b, R_j) < T_{out}(P_a, R_j)$$

or the reservation of process P_b has been accepted after R_j has been granted to P_a, in which case it would have violated reservation condition (3.5) with

$$C_j > T_{in}(P_b, R_j) \,.$$

If a process requests the next resource, the logical clock of the resource is set from the starting timestamp of the reservation to the ending timestamp. The clock condition is always satisfied because of reservation condition (3.1).

We can conclude that the clock condition is also satisfied for the events happening on one resource. This means the following: if process P_b is holding to resource R_j, no reservation on resource R_j exists for any process $P_a \in P$ with

$$T_{out}(P_a, R_j) < C_j \,. \tag{3.9}$$

□

3.6 Proof of Absence of Deadlocks by Contradicting the Circular Wait Condition

We already know from section 3.2 the formal description of our system consisting of conveyor modules and boxes as a distributed multi-process system. In section 3.4, we have defined reservation and granting conditions, and in section 3.5, we have shown that a system following these conditions satisfies the clock condition defined by Lamport (1978).

Our approach to prove the absence of deadlocks is the following: We want to prove that a deadlock cannot occur if the system follows the reservation and granting conditions of section 3.4. Therefore, we have chosen a proof by contradiction. As stated in section 2.3.1, a deadlock exists if the following conditions are fulfilled (Tanenbaum and Bos 2015): First, all involved resources are held by a process. Second, the resources cannot be preempted. Third, all involved processes are in state *Hold and Wait*. And fourth, the processes are waiting in a circular chain. The circular wait condition is the only one that can be negated in our system. To prove the absence of deadlocks, we show that the circular wait condition contradicts the described reservation and granting conditions.

We know that a logical clock in our system is only set forward if an event takes place. In case of a deadlock with circular wait condition, no events take place in a set of resources held by processes. Therefore, the corresponding set of logical clocks would remain infinitely in their current logical time[9]. In our proof, we must show that the logical clocks of resources held by process are set forward definitely.

There is a finite number of processes and resources in the system. Thus, there is a finite number of reservations and there is a finite number of timestamps included in the reservations. Let us consider only the resources held by a process. If we can show that, out of this set of resources, the resources with the minimal value as logical time set their logical clock forward definitely, we can conclude that every event of every process will take place at some time because it will be the event related to the resource with the minimal value for the logical clock at some time.

[9] Logical clocks can also remain infinitely in their current logical time if no process wants to use the resource any more. Thus, the reverse statement is not true.

Proof. Let us assume the worst case: All processes in the system are in state *Hold and Wait*. Let C_{min} denote the minimum logical time of all resources held by a process and

$$R_{min} = \{R_i \mid C_i = C_{min}\}$$

the set of the held resources having the minimal logical time. Processes are only in *Hold and Wait* if granting conditions are not fulfilled. The granting conditions defined in section 3.4.3 must be checked for these processes:

- **GC1** is fulfilled because the processes have already requested the next resource.
- Because of (3.8) and (3.9), there is no resource $R_i \in R$ keeping a reservation for a process $P_a \in P$ with $T_{out}(P_a, R_i) < C_{min}$. Thus, **GC2** is fulfilled for the set of resources R_{min}.
- Therefore, the processes holding these resources can only be in *Hold and Wait* because of **GC3** meaning that the requested resource is not released by the previous process.

Suppose that the processes holding to the resources R_{min} are waiting in a chain as shown in Figure 3.17. In our system, the chain should include at least four different resources so that a deadlock in a loop is possible. Opposing deadlocks with only two resources are prevented by reservation condition (3.3).

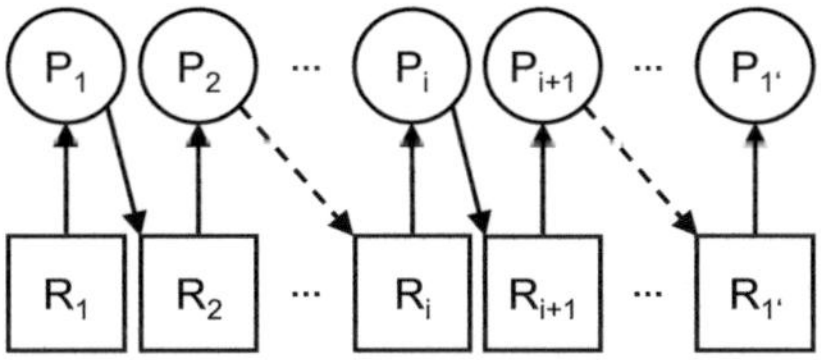

Figure 3.17: Chain of processes in *Hold and Wait*

It is not defined yet whether tandem transport is possible between these resources. From reservation conditions (3.2), (3.3) and (3.4), we can deduce

$$T_{out}(P_i, R_i) = T_{in}(P_i, R_{i+1}) \geq T_{out}(P_{i+1}, R_{i+1})$$

and consequently

$$T_{out}(P_1, R_1) \geq T_{out}(P_{1'}, R_{1'}) \ . \tag{3.10}$$

For there to be a deadlock with circular waiting of these resources, it must be given $P_1 = P_{1'}$ and $R_1 = R_{1'}$. We can deduce

$$C_1 = T_{out}(P_1, R_1) = T_{out}(P_{1'}, R_{1'}) = C_{1'} \tag{3.11}$$

Equation (3.10) is only consistent to (3.11) if tandem transport with reservation condition 3.4 is used exclusively in (3.10). For that, all trios of resources in this circular chain of *Hold and Wait* must satisfy the physical tandem transport condition by being aligned. This is only the case if all resources are aligned, which makes it impossible that $R_1 = R_{1'}$. We can state that a circular waiting of resources connected with tandem transport reservations is physically impossible.

If $R_1 = R_{1'}$, there must be at least one single transport between resources in the chain and equation (3.10) becomes

$$T_{out}(P_1, R_1) > T_{out}(P_{1'}, R_{1'}) \tag{3.12}$$

which is in contradiction to (3.11). A circular waiting of processes is impossible.

□

3.7 Conclusion on the Usage of Logical Time

It is possible to use the principle of logical time in modular material handling systems with decentralized control like GridSorter. By assigning timestamps to transport steps, a partial order between the transport of different boxes is defined during route reservation. When boxes are transported, the conveyor modules must follow the defined sequence in order to prevent deadlocks. The absence of deadlocks in a system using logical time for routing and transport control has been proved. Livelocks are eliminated as well because a box is only allowed to enter the system if a route to its destination has been successfully reserved and because boxes never backtrack on their route.

During transport, starvation cannot occur because the order of accepting processes is defined by the timestamps. Therefore, a certain process is never disadvantaged forever. During the reservation process, starvation occurs if the conveyor modules are not able to find a route for single boxes. In section 4.2.6, we show that the reservation process is complete i.e. finds a valid route for each box.

Since logical time advances in discrete steps only if boxes are transported, the system is robust to variations in transport times. With logical time there is no need to adjust arrival times as it would be necessary using physical time (Mayer 2009). Also, it is not necessary to synchronize the logical clocks of the modules, as would be necessary using physical time. However, it needs to be considered whether the resulting asynchronicity of the logical clocks has an effect on system performance (see section 6.3).

In this chapter, we assume that every transport step is finished after a finite interval in physical time. This assumption is wrong if, for example, the box gets jammed or if the conveyor module breaks down. If the logical time in one conveyor module no longer advances, the entire system can be brought to a standstill. The affected module and its neighboring modules need to be able to detect such a situation and react accordingly, for example by rerouting affected boxes (Fuß et al. 2015; Alt 2014).

Using a routing strategy with time windows, it is possible to consider waiting times due to other boxes during route reservation. The metric to rate different routes is based on lead time (see section 4.2.3). Likewise, the size of the routing decision tree increases as it gains an additional dimension. Consequently, the communication effort for route reservation with time windows is higher than for a route reservation without time windows.

Also, the number of sent messages during route reservation increases with system size which could delay boxes waiting on source modules for the reservation to be completed. There are different ways to deal with this challenge; for example, route reservation could be started a certain time before the box is planned to enter the system. Another approach is to accept partially reserved routes (see section 4.2.5).

Unlike the communication effort for route reservation, the number of messages during the transport of a box decreases in comparison to control algorithms using deadlock avoidance. Because of the predefined order between transported boxes, the granting of a transport step of a specific box only needs two messages representing a clearly limited, local communication effort. The physical transport of a box by far exceeds the duration of this message exchange.

4 Decentralized Control of GridSorter with Logical Time

There are plenty of difficult obstacles in your path.
Don't allow yourself to become one of them.
-- Ralph Marston

In Chapter 3, we applied the principle of logical time to the decentralized control of a material handling system. We showed that the system is free of deadlocks if it fulfills certain reservation and granting conditions. Based on these conditions, control rules need to be defined that each conveyor module has to follow. The challenge is that these local decision rules should create interaction between neighboring conveyor modules that lead to an efficient system behavior. Therefore, the control rules must at least fulfill the reservation and the granting conditions described in section 3.4.2 and 3.4.3.

In the following section 4.1, we first introduce the control architecture and components of one conveyor module. Different components are responsible for different processes.

The reservation manager, for example, is responsible for the reservation process as described in section 4.2 whereas the transport manager is responsible for the transport process as described in section 4.3. Both sections first describe the processes from an observer's point of view and then define the local decision rules.

4.1 Control Architecture and Components

For each box entering the system, a route must be reserved in the so-called reservation process and afterwards the transport process must be performed. Both processes should be controlled by local decisions of conveyors because the system does not possess a central control. Information exchange by message passing is necessary because several conveyor modules are involved in the reservation and transport process of one box. Additionally, several transport processes and reservation processes are running simultaneously because several boxes are in the system.

Each conveyor module has the same control architecture and decision rules because the modules should be identical and interchangeable. Each module can be identified by a unique module ID. To control the different and simultaneously running processes, the control consists of multiple components, each managing different types of tasks and processes. The components, their tasks and their dependencies are shown in Figure 4.1.

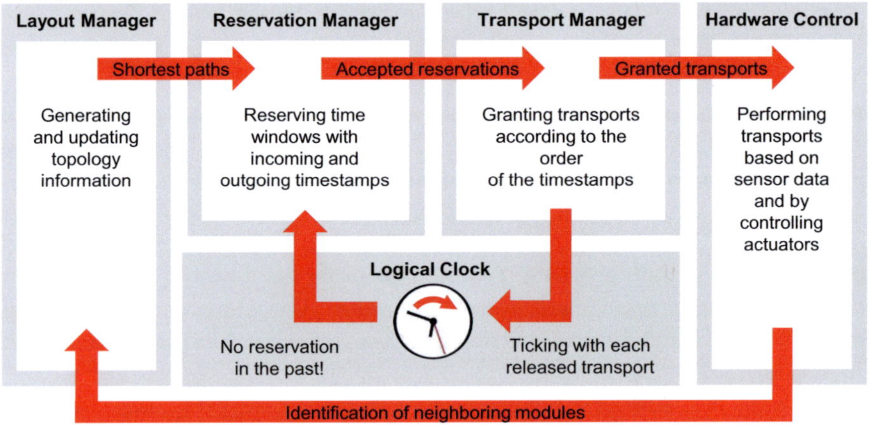

Figure 4.1: Control components and how they influence each other

The **hardware control** is connected to all sensors and actuators. Based on sensor data and in coordination with the predecessor or successor module, it is able to perform the transport of a box from one module to the next. It also controls the electronic connections to the neighboring modules, the so-called FlexPorts. From these connections, it identifies the neighboring modules, which is an important item of information for the layout manager.

The **layout manager** is responsible for the recognition of the topology of the network of conveyor modules. In our system, we use a Link-State-Routing protocol (Tanenbaum 2011). Knowing its own adjacencies, each module sends a message to its neighbors which is then flooded throughout the network. Based on the adjacencies of all conveyor modules, each module calculates for each of its connected ports the shortest paths to all other modules using the Dijkstra algorithm (Dijkstra 1959). Only the minimal path length is stored in the routing table. The algorithms for topology recognition are not within the scope of this thesis because different algorithms have already been tested in simulation by Mayer (2009) and implemented on hardware by flexlog GmbH (2015).

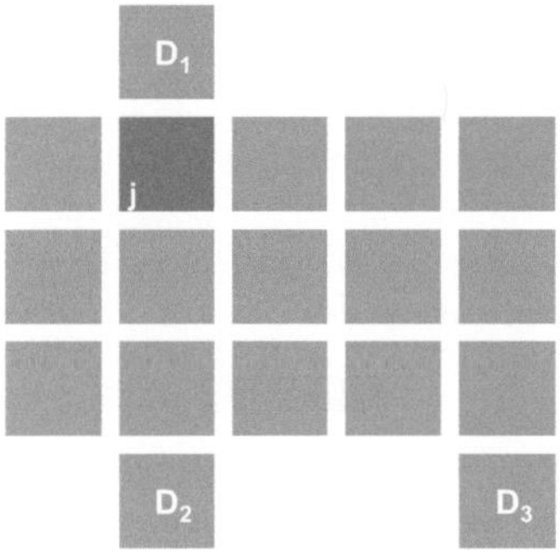

Routing table of module j

Distance to module	*via port* N	S	E	W
D_1	1	∞	∞	∞
D_2	∞	3	5	5
D_3	∞	6	6	8

Figure 4.2: A GridSorter system and the corresponding routing table of module j

Figure 4.2 shows an exemplary topology of a GridSorter system and the corresponding routing table of module j. For simplicity, only the routing table entries with the minimal path length to destination modules are shown. The table content defines the minimum number of necessary transport steps to the destination when using the specified port. Destination D_1 can only be reached via port NORTH whereas the other destinations can be reached via the other three ports.

The inner part of Figure 4.1 looks similar to the control design in Figure 3.9: The dependencies between reservation of modules, transport of boxes and the logical clocks have already been defined in Chapter 3.

Based on the routing table calculated by the layout manager, the **reservation manager** is able to take routing decisions when receiving reservation requests. It decides based on existing reservations whether reservation requests can be accepted. If this is the case, the requests are forwarded to the successor module. Otherwise, they are rejected. These routing decisions are described in detail in section 4.2.4.

The **transport manager** is responsible for the decision whether the transport of a box can be released. It ensures that the transports corresponding to the accepted reservations are performed in the specified order and that no collisions occur. Consequently, it only releases the transport of the earliest planned reservation and hands over this information to the hardware control. The decisions of the transport manager are described in detail in section 4.3.

With each released transport, the **logical clock** is set forward, which then also influences the decisions of the reservation manager because it must not accept a reservation request for past logical time. Each module remains in the time of the transport performed last. Hence, there is no central and no physical time. This difference in logical time can lead to (physical) waiting times. But we believe that in heavily used networks, the modules synchronize automatically with each other, because many transports take place. This proposition is investigated in section 6.3.

4.2 Reservation Process

In Chapter 3 we assumed that a route is successfully reserved for each box before the transport starts. This section presents the reservation process i.e. how the route is reserved. We have several requirements that are already set: The reservation process must execute a time-window-based route reservation because logical time should be used for deadlock prevention. In addition, a decentralized algorithm should be used for route reservation. Since a resource must be reserved before the transport can be requested and performed, it is also required that the reservation process is successfully completed before the first transport step. The reservation process can take place as soon as the destination of the box is known. If the reservation process is not finished before the box arrives on the source module, the box needs to wait on the source module.

What is the objective of the reservation process besides reserving a valid route to the destination? In order to increase system throughput, the travel time of a box should be minimized and the number of boxes in the system maximized according to Little's law $\lambda = N/t$ (Little 1961). Thus, the first objective of the reservation is to find an available path to its specific destination that is short in terms of travel time. The second objective of the reservation is to clear the source module as soon as possible so that another box can enter the system.

After we introduce existing search algorithms for path finding in section 4.2.1, we reason our choice to use iterative deepening A* search for route planning in section 4.2.2. We then present in section 4.2.3 our distributed design of iterative deepening A* search. This section describes the algorithm from the perspective of an external observer, whereas the following section 4.2.4 presents the algorithm from the perspective of one conveyor module: It is explained how a conveyor module takes routing decisions when receiving a reservation message.

One challenge is the scalability of the reservation process. Scalability should be given with regard to the size of the network, the number of boxes in the system and also the number of simultaneously running reservation processes. Depending on these parameters, the duration of the path search can increase drastically. In section 4.2.5, we introduce methods by which the complexity of the search can be reduced in order to accelerate the path search and to assure scalability in the above-mentioned parameters. Finally, in section 4.2.6 the completeness of the route reservation process is proven.

4.2.1 Existing Search Algorithms for Path Finding

In this section, the basic principles and common characteristics of existing path-finding algorithms are presented. For detailed explanations of the search algorithms and information about their algorithmic complexity and time requirements, we refer the reader to the stated literature. Figure 4.3 shows a basic categorization of existing path-finding algorithms and illustrates how characteristics are inherited.

The search for the shortest path is a known problem in graph theory. Usually, the path-finding algorithms sequentially explore nodes to find an optimal path to the destination. The neighboring nodes of the active node are added to the list of open nodes, called *open-list*. The path-finding algorithms differ in the manner in which the node to be explored next is selected from the open-list. All of these algorithms are centralized algorithms, where the nodes are visited sequentially and all information about visited nodes is available.

As uninformed algorithms, the *depth-first search* and the *breadth-first search* (Cormen 2009) are the base for path-finding algorithms. The breadth-first search generates a tree of partial paths and explores all nodes in one level of depth before exploring deeper nodes, whereas the depth-search explores one path in its total depth before backtracking and exploring neighboring nodes closer to the root node. Depth-first is not complete, i.e. it does not terminate if the graph is infinitely large and is not optimal if terminated upon finding the destination. The advantage of depth-first search is the low memory usage.

The *depth-first iterative deepening search* combines the advantages of depth-first and breadth-first search (Korf 1985a) by repeatedly performing depth-first searches with limited depth. With each iteration, the depth limit is increased by one. In this way, the nodes are visited in order of a depth-first search but the cumulative order of newly explored nodes corresponds to breadth-first search. It seems to be wasteful that nodes are visited multiple times, but this disadvantage of a constantly increased run time is negligible compared to the advantage of low memory usage (Korf 1985a).

The *Dijkstra search* (Dijkstra 1959) and *A*-search* (Hart et al. 1968) generate the open-list like the breadth-first search. As best-first searches, they use cost information to select the node to be explored next. Dijkstra uses the path costs from the root node to the open node. A*, on the other hand, uses a predictive heuristic by adding an estimation of the remaining path costs to destination. If the heuristic is admissible, i.e. underestimating the real costs, both algorithms are proven to be optimal in graphs with non-negative path costs. However, they still feature the same disadvantage of high memory usage as the breadth-first search.

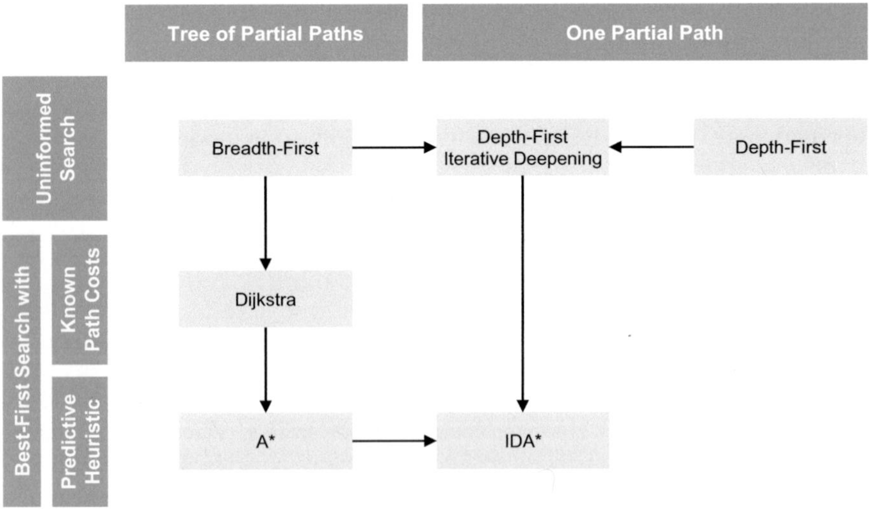

Figure 4.3: Basic categorization of existing path-finding algorithms

The *Iterative deepening A* search (IDA*)* is a combination of iterative deepening depth-first search with the A* heuristic (Korf 1985b). The path costs, estimated as in A*, are used to select the node to be explored next. These heuristic costs are also used to define the depth limit which is iteratively increased. If the depth limit is increased according to the heuristic cost estimation of the root node and the heuristic is admissible i.e. underestimating the true path costs, IDA* finds optimal paths.

4.2.2 Choice of Search Algorithm for Decentralized, Parallel Route Planning

For decentralized route planning, the search for the route path of a box should be performed by the collectivity of modules. Each module hosts its corresponding node in the search graph and knows the predecessor node and potential successor nodes. Nodes are then explored by handing over the search from module to module with the help of messages. Since conveyor modules are only able to send messages to neighboring modules, the search should follow the depth-first principle. Since the route planning should be time-window-based, the search tree is infinitely large due to the temporal dimension. Therefore, the search depth should be limited like in depth-first iterative deepening.

As described in section 4.1, each module knows the minimal path length to a specific destination via a specific port. With this information, the modules are able to estimate the remaining path cost to the destination of a box as a predictive heuristic as performed within A* search. Based on these two characteristics of our system, a decentralized version of IDA* seems to be suitable for the route planning.

In our case, the system has multiple sources for entering boxes. Thus, multiple searches for route planning should be executed in parallel. Each search only requires one conveyor module at a time to host the active node. Consequently, parallel execution of multiple searches is possible because there are multiple conveyor modules in the system. Each module keeps the information about the searches it has participated in and is able to handle multiple searches exploring itself independently of searches exploring other nodes. The influence of the interference of parallel searches on the quality of the route planning cannot be predicted. One might suppose that it increases with the number of active searches in the system. In accordance with the depth-first principle, at most one partial and then total path is reserved for one box simultaneously, which reduces the probability of interference with other searches. The advantage of low memory usage in centralized execution corresponds to low probability of interference in decentralized execution.

4.2.3 Parallel Route Reservation with Decentralized Iterative Deepening A*-Search

The path search algorithm *Iterative Deepening A* (IDA*)* by Korf (1985b) is a search algorithm for finding a time-independent path. The search is executed by a central control. Within IDA*, the search tree is explored depth-first. The node to be explored next is selected from the successors of the currently explored node based on a heuristic that estimates the cost of the total path as within A*. The depth is limited by a maximal path cost and is iteratively increased if the search does not terminate with the current limitation.

The first adaptation to our system is to extend the time-independent path finding to a time-dependent path finding. The search tree becomes a *spatio-temporal search tree.* We use the lead time i.e. the travel time of a box from source to destination as *heuristic costs* because it takes the spatial and temporal dimension into account. The *search depth* is limited by limiting the heuristic costs of a path.

The second adaptation to our system is necessary because the search should be executed on decentralized controls where each conveyor module takes local routing decision. To explore the search tree, the conveyor modules exchange *reservation messages* handing over the search to each other. As an aid to taking reasonable routing decisions while participating in multiple searches in parallel, each conveyor module stores *routing information locally* in a reservation table.

Since the search algorithm IDA* is executed on decentralized controls each of which is hosting a certain set of nodes in the search tree, we call it *Decentralized Iterative Deepening A** in short DIDA*.

The Spatio-Temporal Search Tree

A node in the search tree is not only defined by its location, i.e. its connected conveyor modules. Because of time-dependent route reservation, it is additionally defined by the logical time of the incoming transport of the box. Consequently, the definition of a node in the search tree consists of a spatial and a temporal part. In our case, a node in the search tree is defined by the module ID and the logical timestamp of the incoming transport of the box on this module. The edges between nodes depend on the spatial and the temporal dimensions and represent the reachability of free nodes.

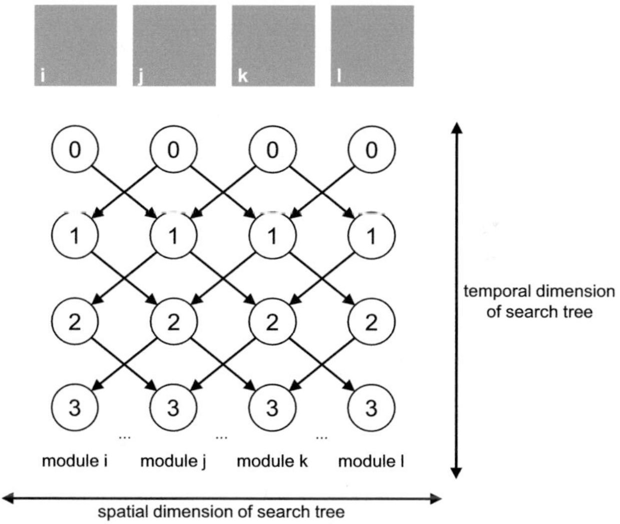

Figure 4.4: The search tree of four modules without any reservation

Figure 4.4 shows an exemplary search tree for a linear conveying line of four modules. The horizontal dimension represents the spatial dimension of the search tree, whereas the vertical dimension represents the temporal dimension. Each node is connected to exactly one node of each neighboring module. The timestamp of a successor node must be greater because of reservation conditions (see section 3.4.2). Message passing is used to explore the search tree and to check for the reachability of free time windows on neighboring modules. If one of the connected nodes is not reachable, it is replaced by a node with a higher timestamp. During the reservation phase, time windows are reserved for a specific box. In accordance with the reservations, the search tree has to be changed by deleting distinct nodes and edges. In Figure 4.5, the first transport step for box *a* has been reserved and the opposing edge has been deleted. The dotted edge indicates that this successor node is reachable but not desirable because the box would backtrack. Figure 4.6 shows the total reserved route for box a. Crossing edges have been deleted to prevent opposing transports whereas parallel edges next to the reserved route are marked as tandem transports because they could take place simultaneously.

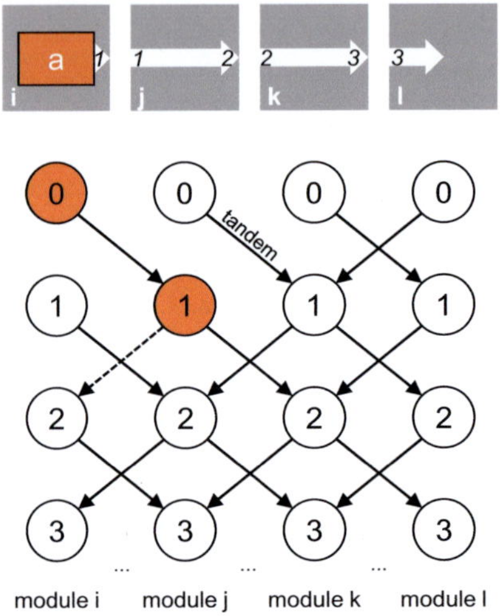

Figure 4.5: The search tree of four modules during the reservation process

If there are no spatial restrictions to routes, an infinite number of possible routes exists for each box because the routes can include an unlimited number of cycles. Consequently, the search tree is infinitely large in the spatial dimension. In order to reduce the size of the search tree in spatial dimension, we identified two methods: One method is to guarantee cycle-free routes[1] by prohibiting boxes from visiting one module more than once. In Figure 4.5, the box is not allowed to move with timestamp *1* from module i to j and in timestamp *2* back again (represented by the dotted edge). Another method to further reduce the size of the search tree is to limit the spatial path length[1] (Alt 2014). One possible implementation is

[1] Restricting the spatial dimension of the search tree also supports that system behavior is perceived as intelligent by humans. Silver (2005) states that cycle-free routes are generally rated as more intelligent and consistent by human observers. We also believe that routes with short path lengths are perceived as more reasonable by human observers because boxes are directly transported to their destination without detours.

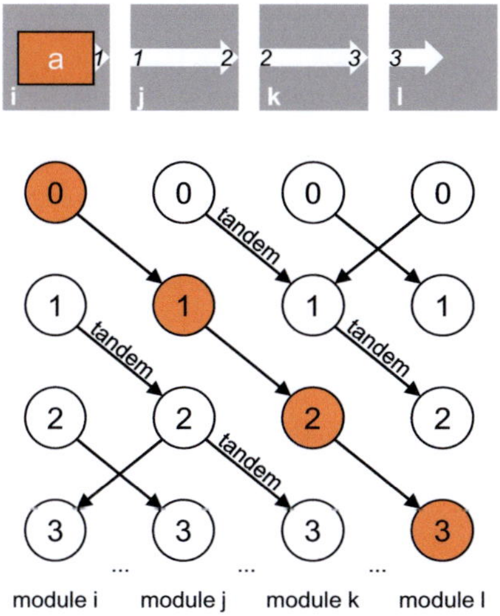

Figure 4.6: The search tree of four modules after the reservation process

to only allow routes with shortest path length (different implementations are presented in 5.3.3).

In addition, the restriction of routes can be defined individually for each module because each module selects the possible successor nodes for each reservation request specifically. Hence, it is possible to define restrictions depending on the position of the conveyor module on the reserved route. For example, it could be interesting that modules close to sources only accept shortest paths whereas modules that are used less accept detours.

Lead Time as Spatio-Temporal Heuristic Costs

The lead time t_{tot} of a path is defined as the travel time of a box from its source to its destination using the specific path. The expected lead time t_{exp} of a path from source to destination is used as heuristic for the selection of the successor node. Like the search tree, the lead time also has a spatial and a temporal part. The lead time consists of the time when the box is moving (spatial part) and the time when the box is waiting (temporal part). When estimating the expected lead time, the spatial part can be deduced from the path length and the temporal part from the logical timestamps of the reservation. The path length can be taken from the routing table and does not need to be calculated for each reservation request. The expected lead time is estimated in the unit of logical time steps.

Once a node is explored, the expected lead time is calculated for all successor nodes to prepare the decision on which node should be explored next. The expected lead time of the successor nodes consists of the upstream lead time t_{up}, which is the lead time from source to the explored node, and the estimation of the downstream lead time, which is the lead time from the explored node via the specific successor node to the destination. The initial estimation assumes the absence of waiting times and calculates the downstream lead time as the minimal path length to destination $l_{dest}(f_i)$ using the port f_i to the specific successor node. The information about the minimal path length is taken from the routing table.

It is also possible to select the successor node based on other criteria than expected lead time. For example, it could be preferable that certain modules, for example the neighbors of sources, are not occupied by waiting boxes. In this case, the module chooses the successor node that offers the earliest outgoing transport. In section 5.3.3, we present implementations where neighbors of source modules use different criteria to select the direction of outgoing transport.

Limiting the Search Depth

The search depth is limited by the maximal lead time t_{max} defining the maximal heuristic costs of a search node to be explored. It limits the number of explored nodes per iteration and, therefore, influences the number of required iterations to find a valid route to destination. Each search iteration is started by the source module that is responsible for the determination of the depth limit. The depth limit must be increased each time. The more it is increased, the fewer search iterations are needed, but the quality of the found path could decrease because the destination can be found before all paths with better or equal costs are explored. Usually, the source module sets the depth limit to its currently expected lead time.

Exploring the Search Tree by Message Passing

Within our path search, IDA* is not executed by one control but by a set of multiple conveyor modules. Each path search is started by the source module by sending the first reservation message. Only one module at a time is allowed to take routing decisions: the module that is hosting the currently explored node in the search tree. The search is handed over to a neighboring module by message passing. If the search tree can be further explored, the module selects a successor node. If the search needs to be backtracked, the search is handed over to the predecessor node. Consequently, the search tree is explored by passing messages from module to module.

Each module only knows its own reservations. Reachability from its own free time windows to free time windows of neighboring modules must be checked by messages. For the exploration of the search tree, we introduce three different types of reservation messages:

- An Explore message is sent if the active module hands over the search to a successor module to deepen the search by exploring the search tree. This is also called a reservation request.
- A Backtrack message is sent if the active module wants to backtrack the search to the predecessor module. The reservation request is rejected.
- A Confirm message is sent to the predecessor module to confirm the reservation because the search has terminated successfully.

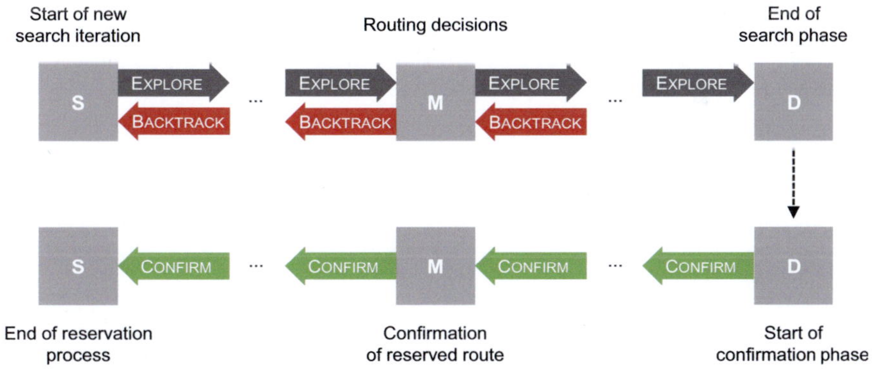

Figure 4.7: Course of reservation: search phase (upper half) and confirmation phase (lower half) whereby S = Source, M = Module and D = Destination

The reservation process can be divided into two phases: the search phase and the confirmation phase. The search phase is started by the source module by sending an EXPLORE messages as shown in Figure 4.7. During the search phase, each module takes routing decisions when receiving an EXPLORE or BACKTRACK message. It then hands over the search to the predecessor module with a BACKTRACK message or to the successor module with an EXPLORE message. If the search is backtracked to the source module, it starts a new search iteration with increased depth limit. Once the search reaches the destination module with an EXPLORE message, the search phase is finished and the confirmation phase starts: Each module on the reserved route gets a CONFIRM message and forwards it to its predecessor. In this way, the reserved route is confirmed from destination to source.

Content for Reservation Message	**Comment**
Reservation ID Destination ID Source ID Maximal lead time t_{max}	General information set by source
Message type {EXPLORE, BACKTRACK, CONFIRM} Timestamp T_{msg} Expected lead time t_{exp}	Updated by each module before sending the message
Port permission Proposed timestamp T_{prop}	Indicating the BACKTRACK cause

Table 4.1: Content of reservation (rsv) message (msg)

A reservation message includes all information that a module needs to take routing decisions for the path search. The general information is set by the source module, node information is updated by each module as is the information indicating the BACKTRACK cause.

Each path search can be identified by a unique reservation ID because several reservation processes are running in parallel. The reservation ID is set by the root node i.e. the source module, and consists of the source module ID and the logical timestamp of the incoming transport of the box on the source module. When the source module initializes a path search it includes general information in the first reservation message, which is then included in all forwarded messages. This general information of one reservation includes the reservation ID, the destination module ID, the source module ID and the current maximal lead time t_{max}.

In addition to the general reservation information, a reservation message includes the following information which is updated by the module sending the message: the type of reservation message, the logical timestamp T_{msg} of the transport between the sending and the receiving module and the expected lead time t_{exp}.

The search can be backtracked because of two different causes that are indicated in the BACKTRACK message:

1. If the branch behind this node is fully explored, the reservation request should not be sent again during the current search iteration. Consequently, the module sends a BACKTRACK message denying port permission. This means that the node can only be explored in subsequent search iterations.
2. If the path with the requested timestamp ends in a dead end because no free time window is reachable from this node due to other reservations, the module proposes an alternative timestamp T_{prop}. This means that another free time window should be used. With the information of the BACKTRACK message, the predecessor module deletes the node with the dead end from the search tree and substitutes it with another successor node using the proposed timestamp T_{prop}. If the expected lead time of the new node exceeds the maximal lead time, the module additionally denies port permission because the search depth of the current search iteration would be exceeded. In this case, the new node can only be explored in subsequent search iterations.

Figure 4.8 shows four conveyor modules where only module k holds a reservation. The sequence of messages is shown in the upper part, while the explored search tree is shown in the lower part. The first search iteration is started by the source module with a depth limit of *3* as maximal lead time since the minimal path length to the destination module is *3*. When the search reaches module k, it replies with a BACKTRACK message because the reservation request interferes with the existing reservation. Module k proposes timestamp *4* as alternative. This alternative timestamp indicates that the box would have to wait on module j: the resulting estimated lead time exceeds the depth limit. Therefore, module j sends a BACKTRACK message denying permission to send the reservation request for the proposed timestamp during this search iteration. The source module starts another search iteration with depth limit *5* which is successfully completed.

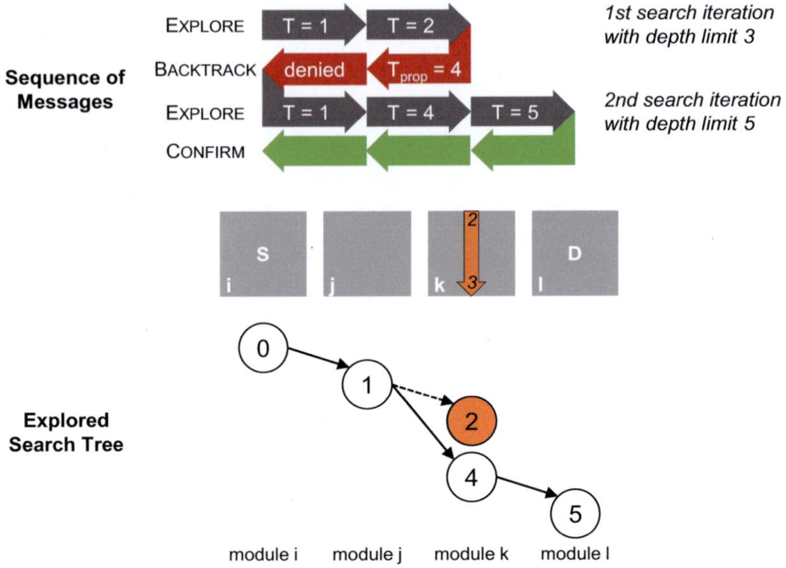

Figure 4.8: Example for the exploration of a search tree by message passing

Storing Routing Information Locally

In systems with decentralized control, each module must store the relevant information of the explored nodes it is hosting. Therefore, it keeps the information of explored nodes in two different reservation tables. Accepted reservations are kept in the active reservation table. Rejected reservations are kept in the inactive reservation table. In central implementations of IDA*, the information about explored nodes is deleted after backtracking. This is why one of its main advantages is low amount of memory required. Thanks to decentralized control, the information about explored nodes can be kept in order to shorten the subsequent search iterations. The information is deleted once it becomes irrelevant i.e. if the logical clock is set forward in such a way that the inactive reservation entry lies in the past.

In addition to the general reservation information, a reservation entry includes the following information:

- The state of the reservation table entry: EXPLORE and CONFIRM reservation entries are kept in the active reservation table. Entries with state BACKTRACK are kept in the inactive reservation table.

- The description of the reservation: The start of a reservation is defined by the incoming port f_{in} and the timestamp of the incoming transport T_{in}. The end of a reservation is defined by the outgoing port f_{out} and the timestamp of the outgoing transport T_{out}.
- The routing information: For each reservation table entry, the upstream lead time t_{up} is stored. Also, the expected lead time $t_{exp}(f_i)$, the outgoing timestamps $T_{out}(f_i)$ and the permission are stored separately for each connected port f_i.

The routing information is used to take routing decisions i.e. decisions as to which successor node an Explore message should be sent or if the reservation should be rejected. It is initialized with the reception of the first Explore message, updated with information of Backtrack messages and partially reset in subsequent search iterations (compare section 4.2.4).

An overview of all information included in a reservation message and a reservation table entry can be found in appendix A.

4.2.4 Local Routing Decisions during Search Phase

In the previous section, we explained how the search algorithm is designed in order to find a time-dependent route with decentralized control. The data formats of reservation messages and reservation table entries have been introduced. In this section, the routing decisions are explained in detail, i.e. we describe how a conveyor module reacts if it receives a reservation message. A first version of local routing decisions has been developed together with Oberkersch (2013).

Figure 4.9 shows a combination of activity charts of the search phase. Activities are represented as rectangles and decisions as diamonds. Three events serve as starting points (left side of chart): the start of a reservation process by the source module at the beginning of the search phase and the reception of an Explore or Backtrack message during the search phase. All four ending points of the chart (right side of chart) are defined by sending of a message: Whereas sending an Explore or Backtrack message hands over the search to another module, the Confirm message indicates the end of the search phase and the beginning of the confirmation phase. The gray backgrounds highlight the three main activities: the decision on incoming transport, the generating and updating of routing information and the decision on outgoing transport. All three main activities are explained in detail in the following sections.

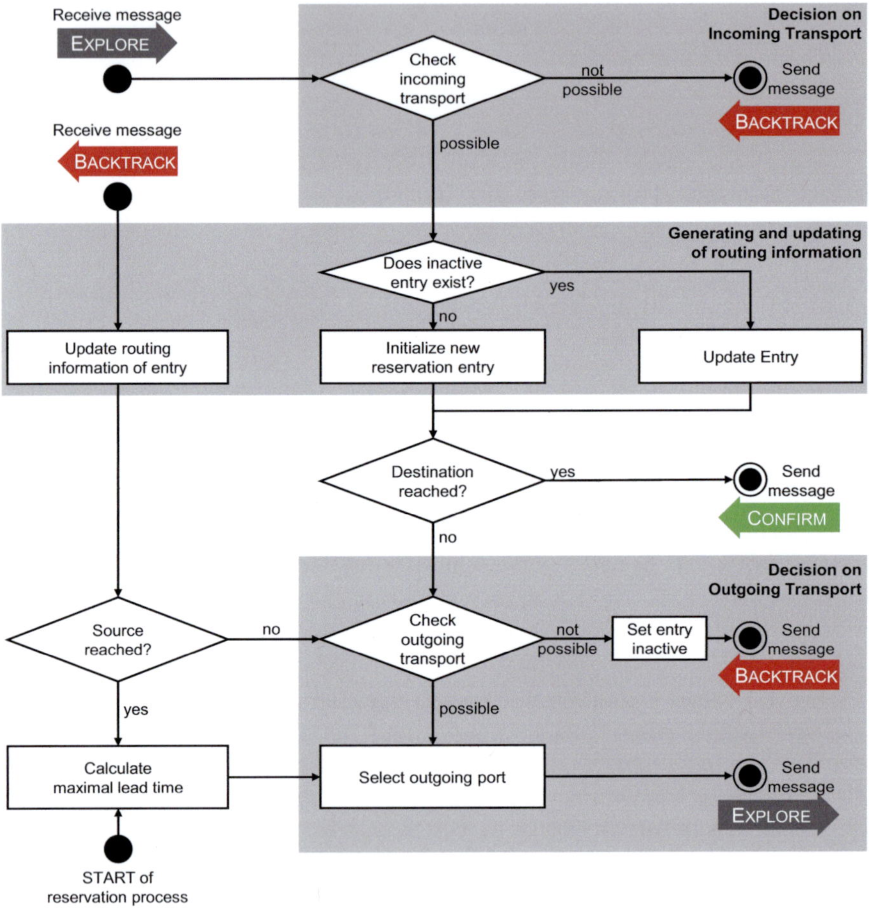

Figure 4.9: Combined activity chart for events during search phase

Let us go through Figure 4.9: If a module receives an EXPLORE message, it first checks whether the requested incoming transport can be performed through this incoming port at the specific timestamp. If the incoming transport is not possible, it directly sends a BACKTRACK message. If, on the other hand, the incoming transport is possible, the module generates or updates the routing information. The routing information is also updated if a module receives a BACKTRACK message. Then, the module checks whether an outgoing transport is possible depending on the routing information and the other entries in the active reservation table. If the outgoing transport is possible, the module sends an EXPLORE message in the most promising direction. If, on the other hand, the outgoing transport is not possible, the module sends a BACKTRACK message to the predecessor module. The source and destination modules react differently: If a source module receives a BACKTRACK message, it increases the depth limit to send another EXPLORE message. If a destination module receives an EXPLORE message, it starts the confirmation phase by sending a CONFIRM message.

Decision on Incoming Transport

The decision whether or not an incoming transport is possible depends on the entries in the active reservation table. We identified four different reasons why a request should not be accepted because of the incoming transport.

Requested timestamp in the past If the timestamp of the requested incoming transport is lower than the logical clock of the module, the reservation request should be rejected in order to guarantee advancing logical time (see 3.4.2). In this case, the module includes the following information in the BACKTRACK message: It proposes the earliest timestamp T_{prop} higher than the logical clock and not colliding with another reservation. It increases the expected lead time t_{exp} by the difference between this proposed timestamp and its own logical clock C.

$$t_{exp} := t_{exp} + T_{prop} - C$$

If the expected lead time exceeds the maximal lead time, the depth limit of this search iteration has been reached. Therefore, the module denies permission to send the reservation request for the proposed timestamp during this search iteration.

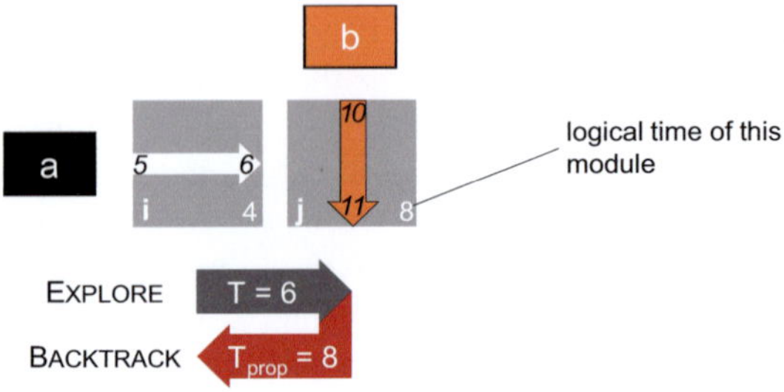

Figure 4.10: Example for rejection because of requested timestamp in the past

Figure 4.10 shows a situation in which a reservation request is rejected because the requested timestamp for the transport lies in the past. The logical clock of module i is set to *4* whereas the logical clock of module j is set to *8*. Once module i requests a transport for timestamp *6*, module j sends a BACKTRACK message proposing timestamp *8* for the transport. The other reservation for box b on module j is not interfering.

Reservation sent in a loop If there is another reservation entry for the same reservation, the reservation has been sent in a loop. Within the BACKTRACK message, the module denies permission to send the reservation request again during this search iteration. Independent from the expected lead time, the module indicates that the search tree should not be further explored to avoid a loop in the route.

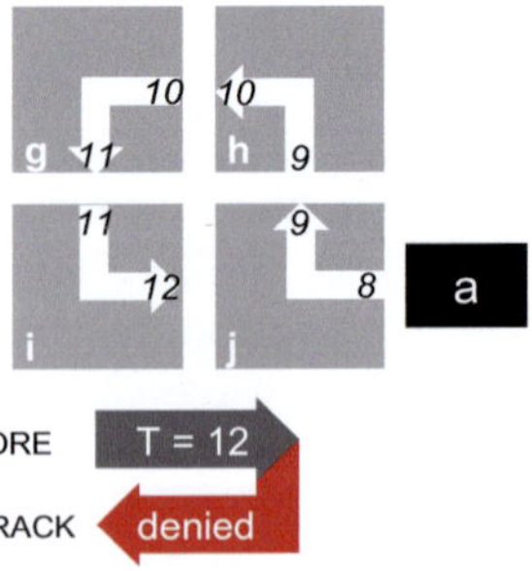

Figure 4.11: Example for rejection because of reservation sent in a loop

Figure 4.11 shows a situation in which a reservation is sent in a loop. Since module j already holds a reservation for box a, it is not allowed to accept a second one. Within the BACKTRACK message it denies permission to send the reservation request again. Thus, module i now needs to try another port or has to backtrack the search to module g.

Crossed reservation requests If two adjacent modules simultaneously send reservation requests for the same timestamp to each other, they could be caught in a closed loop raising the sending timestamps infinitely. If this situation is detected based on the existing reservation table entries, one of the modules first rejects the reservation (without permission to resend it) and then sends its own reservation request again. BACKTRACK and EXPLORE messages have to be sent in this order because the reservation table of each module has to be consistent at all times, meaning no two accepted reservations should interfere with each other.

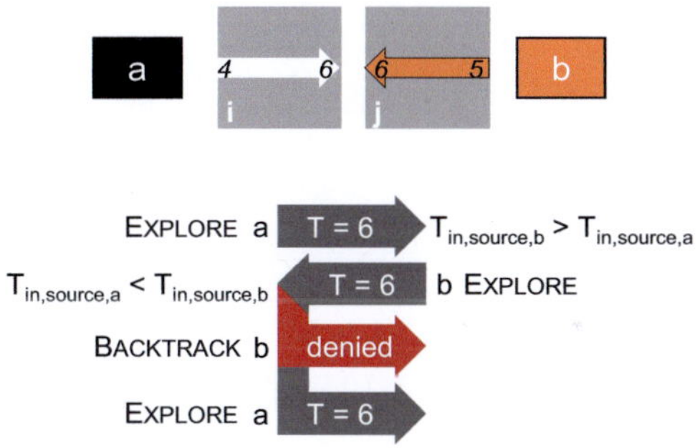

Figure 4.12: Example for rejection because of crossed reservation requests

Figure 4.12 shows the example for crossed reservation requests. Module i and j simultaneously send reservation requests including timestamp *6* to each other. Both detect the situation and compare the reservation IDs of the reservation requests. They first compare the incoming timestamp of the source and then the source module ID. Since the reservation of box a is "older", module i sends a BACKTRACK message for box b and resends the EXPLORE message for box a.

Interfering with another reservation If the timestamp of the requested incoming transport collides with another reservation, i.e. the module is not available at that timestamp, the module rejects the reservation. It includes the information when it is available again and increases the expected lead time by the difference between the newly proposed timestamp T_{prop} and the previously requested timestamp T_{msg} included in the EXPLORE message.

$$t_{exp} := t_{exp} + T_{prop} - T_{msg}$$

If the expected lead time exceeds the maximal lead time, it denies permission to send the reservation request for the proposed timestamp during this search iteration.

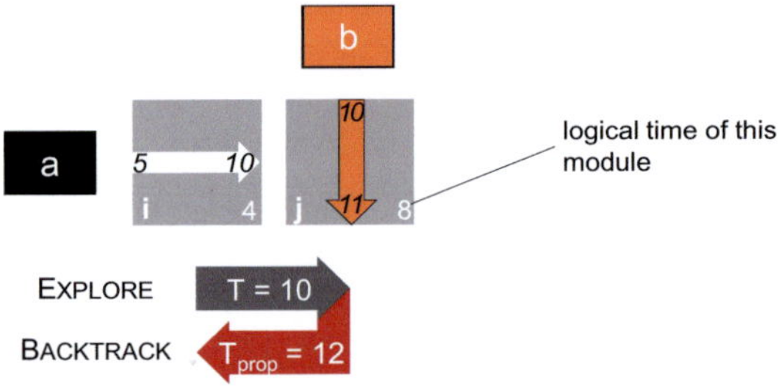

Figure 4.13: Example for rejection because of another reservation interfering with the incoming transport

In Figure 4.13, module j rejects the reservation request for box a because it interferes with the existing reservation for box b. It proposes *12* as an alternative timestamp because it is the first available timestamp after the reservation for box b.

If the incoming transport is possible, the routing information is generated or updated. If destination is reached, the module replies with a CONFIRM message. Otherwise, it must take the decision on outgoing transport.

Decision on Outgoing Transport

The decision whether and in which direction an EXPLORE message can be sent, depends on the routing information included in the relevant reservation table entry and the other entries in the active reservation table.

The restrictions of routes can be implemented by defining a set of eligible ports from which a conveyor module can select the outgoing port i.e. the successor module. Let F_{all} be the set of all connected ports of one module.

$F_{elg} \subseteq F_{all}$ is then the set of eligible ports for a specific reservation from which the outgoing port can be selected. If the routes are not restricted, $F_{elg} = F_{all}$ is valid.

Basically, we identified two different reasons why a reservation should be rejected because of the outgoing transport.

All ports denied If all modules in the direction of eligible ports have sent a BACKTRACK message with denial, there is no longer any permitted path and the module sends a BACKTRACK to the previous module without permission to send another reservation request in the current search iteration. It includes in the message the minimal expected lead time of all eligible outgoing ports because it is relevant for the subsequent search information.

$$t_{exp} := \min_{f_i \in F_{elg}} (t_{exp}(f_i))$$

Rsv table of module j

Box	*[T_{in}, T_{out}]*	*[f_{in}, f_{out}]*	t_{max}		**N**	**S**	**E**	**W**
				T_{out}	6	10	8	8
a	[5 , -]	[N , -]	20	t_{exp}	14	16	17	21
				perm	in	den.	den.	den.

Figure 4.14: Example for rejection because of all ports denied

Figure 4.14 shows an exemplary reservation table entry for box a. T_{out} and f_{out} are not defined yet because no outgoing port has been selected. NORTH cannot be selected as the outgoing port because it is already the incoming port. From all other directions, the module has received BACKTRACK messages denying permission. Module j must also send a BACKTRACK to NORTH denying permission.

Interfering with another reservation If the outgoing timestamps of all permitted directions interfere with the next reservation in the active reservation table, the module sends a BACKTRACK message. It includes the information when it is available after the next reservation and increases the expected lead time accordingly. If this calculated expected lead time exceeds the maximal lead time, it denies permission to send the reservation request for the proposed timestamp.

$$t_{exp} := t_{up} + T_{prop} - T_{in} + \min_{f_i \in F_{elg}} l_{dest}(f_i)$$

Rsv table of module j

Box	*[T_{in}, T_{out}]*	*[f_{in}, f_{out}]*	t_{max}		**N**	**S**	**E**	**W**
				T_{out}	6	10	8	8
a	[5 , -]	[N , -]	20	t_{exp}	14	16	17	21
				perm	in	-	-	den.
				T_{out}	8	**10**	8	8
b	[7 , 10]	[N , S]	12	t_{exp}	12	**12**	14	16
				perm	in	-	den.	den.

Figure 4.15: Example for rejection because of another reservation interfering with the outgoing transport

Figure 4.15 shows an example in which no outgoing port can be selected because of another interfering reservation. The reservation table shows two entries; an outgoing port should be selected for the entry of box *a*. Again, NORTH cannot be selected because it is the incoming port. WEST has been denied because the maximal lead time has been exceeded. SOUTH and EAST could be selected according to the port permission but the outgoing timestamps interfere with the reservation of box *b* with incoming timestamp *7*.

If there is at least one permitted direction whose outgoing timestamp does not collide with the next reservation, the module needs to select the outgoing port in order to send an EXPLORE message in this direction.

Selecting the Outgoing Port In order to meet the first objective of the reservation process to find routes that are short in terms of travel time of the box, the module usually selects the port with the lowest expected lead time out of the possible ports F_{elg}. If there are several ports with the same rating, it prefers the port opposite to the incoming port because direction changes are time-consuming. If both ports with direction change are rated equivalent, they are selected alternately.

The criteria for selecting the outgoing port can be different depending on the position of the module. The source module and its neighbors use criteria differing from the usual criteria in order to free the source module as soon as possible: These modules select the outgoing port with the lowest outgoing timestamp independently of the expected lead time.

Once the outgoing port is selected, the information for the EXPLORE message is determined:

$$T_{msg} := T_{out}(f_{out})$$
$$t_{exp} := t_{exp}(f_{out})$$

Rsv table of module j

Box	*[T_{in}, T_{out}]*	*[f_{in}, f_{out}]*	*t_{max}*		**N**	**S**	**E**	**W**
				T_{out}	6	10	8	8
a	[5 , -]	[N , -]	20	*t_{exp}*	14	16	17	21
				perm	in	-	-	den.

Figure 4.16: Example for selection of outgoing port

In Figure 4.16, SOUTH and EAST can be selected as outgoing ports. With the common selection criteria, usual modules select SOUTH as the outgoing port because it has the lower expected lead time. Modules that are neighbors of sources select EAST as outgoing port because it has the lower outgoing timestamp.

Generating and Updating of Routing Information

The routing information included in the reservation table entry is generated, updated and reset depending on the information of received messages.

As shown in Figure 4.9, the routing information is updated when receiving an EXPLORE or BACKTRACK message. A distinction is made between receiving a reservation request for a search node defined by reservation ID and incoming timestamp for the first time, and receiving it again after having initially rejected it.

Receiving a reservation request for the first time If no inactive reservation entry exists for the same reservation ID and the same incoming timestamp, the routing information has to be generated. The outgoing timestamp for every port is set one timestamp later than the incoming timestamp. The upstream lead time for this request and the expected lead time for all ports are calculated based on the message information and the path length information provided by the layout manager (compare section 4.1). The permission is set to default state: each port is permitted for outgoing transport.

$$\begin{aligned} t_{up} &:= t_{exp} - \min_{f_i \in F_{elg}} l_{dest}(f_i) \\ t_{exp}(f_i) &:= t_{up} + l_{dest}(f_i) \\ T_{out}(f_i) &:= T_{msg} + 1 \end{aligned}$$

Receiving a reservation request again If an inactive entry exists for the same reservation ID and the same incoming timestamp, this entry is reactivated. The incoming port is updated according to the receiving direction of the message. If the maximal lead time of the message is identical to that of the reservation entry, this module has already searched for a path during this search iteration and therefore the routing information is not updated. If, on the other hand, the maximal lead time has changed, the permission is reset to default so that all possible outgoing ports can be chosen again.

Receiving a reservation rejection If a module receives a BACKTRACK message for an active entry, it only updates the routing information for this port: It sets the outgoing timestamp to the proposed timestamp and updates the expected lead time and the permission in accordance with the contents of the message. In addition, it updates the maximal lead time as contained in the message.

4.2.5 Partial Route Reservation

With increasing system size, the search tree and, therefore the duration of the reservation process increases. As a consequence, the boxes would have to wait on the source module because of the long duration of the reservation process. One possibility to achieve scalability with increasing system size is to start the transport of a box while the reservation process is still running. Korf (1990) also encounters the problem that the exponential run time of IDA* is not acceptable in certain real-time applications (e.g. example computer games). He presents one solution called *threshold-limited-IDA** where the decision about the first move is taken after a search iteration without finding the destination based on the heuristic costs of all explored nodes. In our system, this would correspond to starting the transport process when a partial route has been reserved. The idea of reserving partial routes within GridSorter has also been presented by Alt (2014).

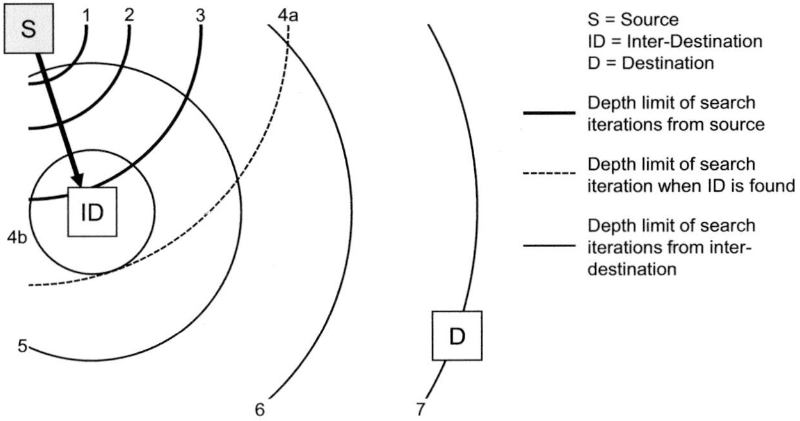

Figure 4.17: Schematic representation of the partial route reservation

What is the method? We want to split the route reservation process into sequential parts. The beginning of the route is fixed while the remaining part of the route to the destination is not found yet. The confirmation of the partial route can basically be initialized by any module keeping a valid, reserved time window of the relevant route. The module is able to fix the upstream route by sending a CONFIRM message and is then acting as a new source for searching the downstream part of the route. We call this module *inter-destination*. Figure 4.17 schematically represents the method of partial route reservation in an example. During the fourth search iteration from the source module, an inter-destination is found. The partial route to this module is confirmed regardless of progress of the search iteration. This search iteration numbered *4a* is substituted by the first search iteration from the inter-destination module numbered *4b*. Neighboring paths to the reserved partial path are not searched any more. In this example, the destination is found during the 7th search iteration while the box is already being transported to the inter-destination module.

What has to be considered in order to prevent deadlocks? Because of the running reservation process, it is not guaranteed that the reservation will be confirmed as planned. Consequently, the inter-destination must not have any other reservations later than the relevant reservation. In this way, it is theoretically able to buffer the box infinitely. Like this, the reservation process can continue without additional restrictions because the inter-destination module can raise the outgoing timestamp with each new search iteration until a valid route to the destination is successfully found.

When should an inter-destination be selected? Selecting an inter-destination module might increase the lead time of the total route because a part of the route is confirmed while it is not known what lead time the total route will achieve. In addition, it is not possible to estimate the potential increase of lead time because no upper limit of lead time for any route is known. Nevertheless, the deeper the search tree has been explored, the better the selection of partial route. Consequently, an inter-destination module should be selected as late in the reservation process as possible and only if necessary. The objective of partial route reservation is to avoid that boxes wait on source modules because of ongoing route reservation. Therefore, the source module decides whether an inter-destination module should be searched based on the progress of the reservation process. If the source module receives a BACKTRACK message indicating that the current search iteration was unsuccessful, and if there is a risk that the new search iteration will exceed the duration of the transport of the box to the source module, the source module enables the search for the inter-destination and includes this information in the reservation message. With the existing routing information of the previous search iterations, the most promising partial routes are explored first. If a module decides to act as inter-destination, it disables the search for another inter-destination by changing the information in the reservation message.

What kind of module is suitable as inter-destination? An inter-destination module is not allowed to accept any reservation requests later than the reservation concerned. To avoid the complete blockage of other reservation processes, neighbor modules of sources or sinks or other inter-destinations are not allowed to be inter-destinations. The objective of choosing an inter-destination is to increase the time which is available for the reservation process before it delays the start of the transport process. Consequently, it should be at a certain distance from source. In addition, the probability that the inter-destination module lies on a good route increases with length of the partial reserved route. We therefore formulate the condition that the remaining distance between inter-destination and destination should not exceed a certain distance. This condition can be parametrized and is defined by a portion of the total distance between source and destination.

The impact of the partial route reservation is studied in section 6.5.

4.2.6 Completeness of the Reservation Process

In this section, we discuss the completeness of the route reservation process. Completeness is given if a valid route is found for every box regardless of the initial situation i.e. the layout, the source module and existing reservations. As described in previous sections, DIDA* is used as path-finding algorithm being executed on decentralized controls. Korf (1985b) has proven IDA* to be optimal, and therefore complete, if the heuristic is admissible i.e. underestimating the real path costs. Because of asynchronous logical clocks, we cannot guarantee that our heuristic is admissible i.e. always underestimating remaining path costs. We can still conclude that DIDA* is complete if it is run as decoupled path search.

However, multiple path searches run in parallel in our system. How do these parallel searches interfere with each other? And how does this impact the completeness of the route reservation process? Because of decentralized control and parallel running path searches, we are not able to prove completeness but we present some arguments suggesting that parallel searches do not compromise completeness.

The interference of parallel searches could have different negative effects: On the one hand, communication livelocks could occur where messages are sent between the same modules repeatedly. To avoid communication livelocks, it is important that with each reservation message, a decision is taken and routing information changed. Situations in which a module sends an identical reservation request again can lead to livelocks because this identical request is declined all over again.

- Each BACKTRACK message either indicates that the request should not be send again or proposes another timestamp. Therefore, each reservation request differs from previously sent reservation requests.

- If a module simultaneously[2] receives two reservation requests that interfere with each other, one reservation message is handled first. The other one is handled second taking into account the reservation entry of the first request.
- If two modules simultaneously send interfering reservation requests to each other (compare to situation *Crossed Reservation Requests* in section 4.2.4), the modules decide which reservation request to accept using the reservation ID. In this way, the path searches are brought into a fixed order regardless of the node where they are interfering with each other. The logical timestamp included in the reservation ID is the first criterion in order to prefer "old" route searches over "new" ones. If both logical timestamps are identical, the source module ID is used as criterion.

We claim that these methods successfully avoid communication livelocks. On the other hand, reserved routes could block each other in such a way that no search is terminated successfully. Because of a finite number of reservations in the system, the system is empty at some point in the future. Hence, a route can be found more easily the higher the timestamps. How is the interference of reservation requests handled by the conveyor modules?

- If a reservation request interferes with other reservations, a later timestamp is proposed. Like this, this path search is "pushed into the future" where the system is more likely to be free of other reservations.
- The information about interference is kept for the next search iteration, even though the interfering reservation might not exist anymore because another route has been reserved. Like this, we avoid that each search iteration explores identical routes again. By contrast, the path search is further "pushed into the future" where the system is more likely to be free of other reservations.

We claim that these methods successfully handle interfering reservation requests in such a way that one of the path searches can continue to find a path whereas the other path search generates reservation requests for later timestamps. Simulation of the system supports our claim (see Chapter 6).

[2] Since the reservation manager is single-threaded, two simultaneous events have to be handled one after another. Usually, the order is randomly chosen.

4.3 Transport Process

Based on the granting conditions for the transport of a box as introduced in section 3.4.3, we now present decision rules for the decentralized control of the transport process of one box. The transport manager introduced in section 4.1 is responsible for the decisions described in the following sections.

The transport process is only allowed to be started after the reservation process is successfully finished. Each module stores the information of the relevant reservation locally. Another requirement is the ability of the hardware control to execute the transport of a box. For the transport of a box from one module to another, both hardware controls have to be notified about the transport of the box. They then communicate with each other in order to execute the transport as soon as it is physically possible. The hardware control is also able to execute a sequence of different transport steps in the order of receipt of the transport notification.

4.3.1 Local Coordination of Transport Steps

We first explain the message types used for the coordination of a transport step before explaining the events and actions that are performed by a conveyor module during the execution of the incoming and the outgoing transport of one reservation entry.

Message Passing for Coordination

The coordination of a transport step is limited to the two participating modules i.e. the sending module and the receiving module. As shown in Figure 4.18, the transport manager uses two different transport message types: REQUEST and GRANT. The sending module sends a REQUEST message to the receiving module indicating that it seeks to perform the transport of a box. Once the receiving module replies with a GRANT message, both modules notify their hardware control about the transport.

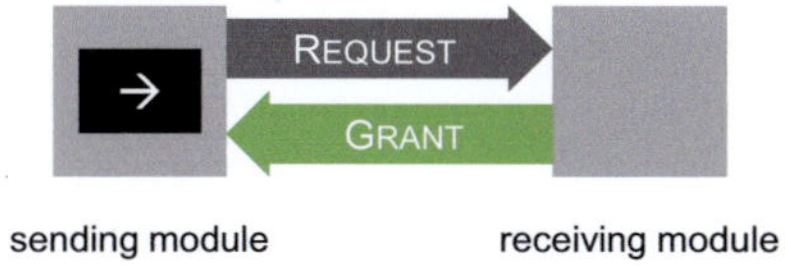

Figure 4.18: The two transport message types

With the content of the transport message, the modules verify that they both hold a corresponding reservation table entry containing the same information about the transport step. A transport message consists of the message type, the reservation ID and the logical timestamp of the transport step that should be performed.

An overview of all information included in a transport message can be found in Appendix A.

Execution of Incoming and Outgoing Transport Step of a Reservation Entry

A reservation entry defines two transport steps by setting port and logical timestamp: the incoming transport and the outgoing transport. During the execution of both transports, the transport state indicates which transport should happen next and if it is already confirmed. The transport manager includes this additional information in the reservation entry. There are four different transport states: RECEIVING REQUESTED, SENDING GRANTED, SENDING REQUESTED and SENDING GRANTED. We explain the meaning of these transport states by describing the execution of both transport steps of one reservation (see Figure 4.19). The position of the box and the sent messages are represented graphically in the middle block of the figure. All actions, events and transport states are described for the middle conveyor module. The transport state of the reservation entry is written above the module. Events are listed on the left, whereas triggered actions are listed on the right.

The first event related to the transport process of a reservation entry is the reception of a REQUEST message from the predecessor module. The middle conveyor module updates the transport state to RECEIVING REQUESTED and checks the granting conditions for the incoming transport. Details of checking the granting conditions are explained in the next section 4.3.2.

Once the granting conditions are fulfilled, the module replies to the predecessor with a GRANT message and changes the transport state to RECEIVING GRANTED. In addition, it sets its logical clock to the timestamp of the incoming transport step and notifies its hardware control that the box can be received.

Once the receiving transport has physically started, the hardware control informs the transport manager. The transport manager sends a REQUEST

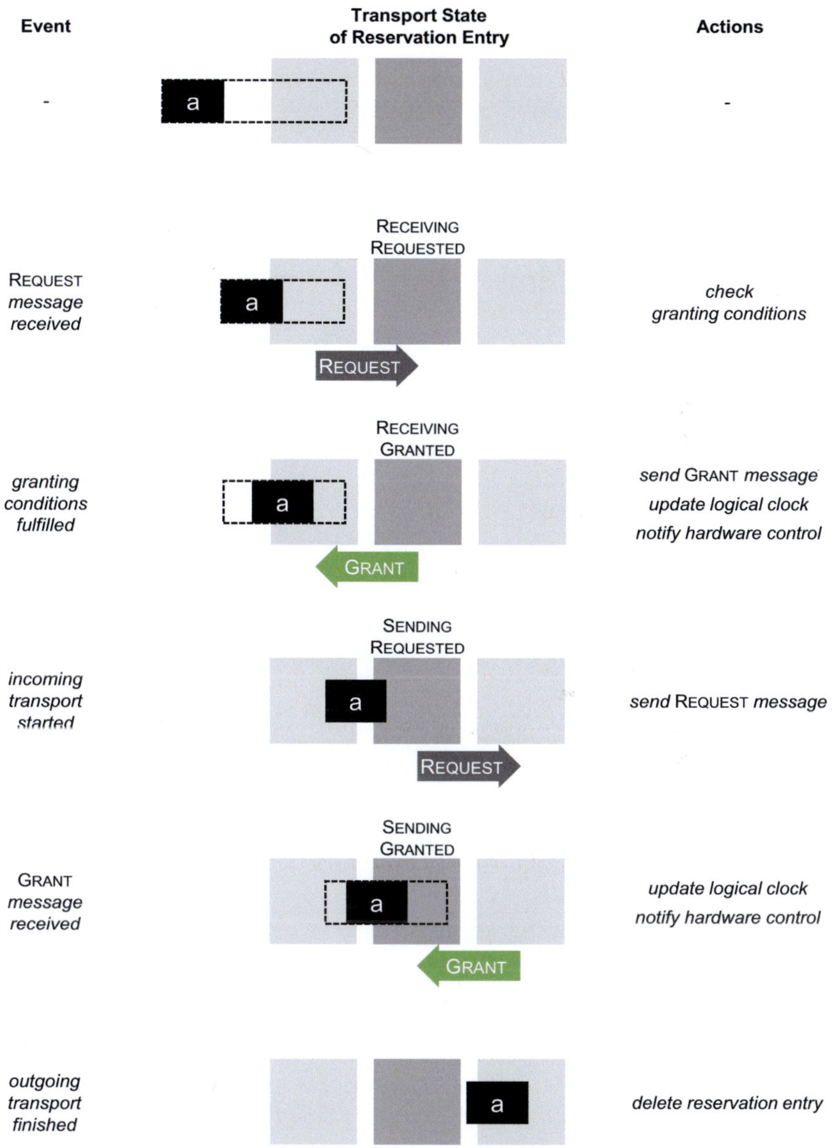

Figure 4.19: Events, transport states and actions during the execution of the transport steps of one reservation

message for the outgoing transport to the successor module and changes the transport state to SENDING REQUESTED accordingly.

On receiving the GRANT message from the successor module, the middle module updates the transport state to SENDING GRANTED. Additionally, it sets its logical clock to the timestamp of the outgoing transport and notifies its hardware control that the box can be sent.

If the hardware control informs the transport manager that the sending transport is finished, the transport manager deletes the reservation table entry. Both transport steps have been executed successfully.

4.3.2 Local Transport Granting Decisions

In the previous section, the execution of transport steps has been described by defining events and triggered actions. In this section, we define conditions that must be fulfilled by a module before granting a transport by sending a GRANT message to the predecessor module. For a correct control algorithm, it is also necessary to define when a module should check the granting conditions.

What are the conditions for granting an incoming transport?

The conditions can be deduced directly from section 3.4.3. **GC1** states that the predecessor module must have requested the transport. Consequently, only reservation entries with transport state RECEIVING REQUESTED are eligible for granting. **GC2** states that only the incoming transport with the lowest timestamp is allowed to be performed. **GC3** is related to physical feasibility of the transport step depending on the status of both modules performing the transport together. This condition is monitored additionally by the hardware control. In section 3.6 it has been proven that the system is deadlock-free. Consequently, the transport step will become physically possible at some time, allowing the hardware control to perform the transport.

Which information is used for checking the granting conditions?

The transport manager is responsible for checking whether the granting conditions are fulfilled. The conditions are checked based on the information in the reservation table. We can assume that the reservation table is ordered by the values of the timestamps of the incoming transport. Since the transport steps have to be performed in the order of their timestamps, it is sufficient to check the first entries:

For single transport, only the first reservation entry is relevant. If the transport state indicates RECEIVING REQUESTED, the incoming transport step is granted by sending a GRANT message to the predecessor module. Physical feasibility is given because the previous transport must have been finished since no previous reservation entry exists[3].

For tandem transport, the second reservation entry is relevant. Granting conditions are fulfilled if the first reservation entry is in transport state SENDING GRANTED and the second entry is planned as tandem transport, i.e. the incoming timestamp corresponds to the outgoing timestamp of the first entry. The receiving module sends a GRANT message to the predecessor module. Physical feasibility is given because the previous box is already being transported to the next conveyor module.

In Figure 4.20, module k is not allowed to grant the incoming transport of box b from module l because another transport of box a is planned with a lower timestamp. It is sufficient to check the reservation entry of box a. The waiting of box b is necessary because the logical clocks of the four modules are not synchronized. The objective is to keep the logical clocks as synchronized as possible.

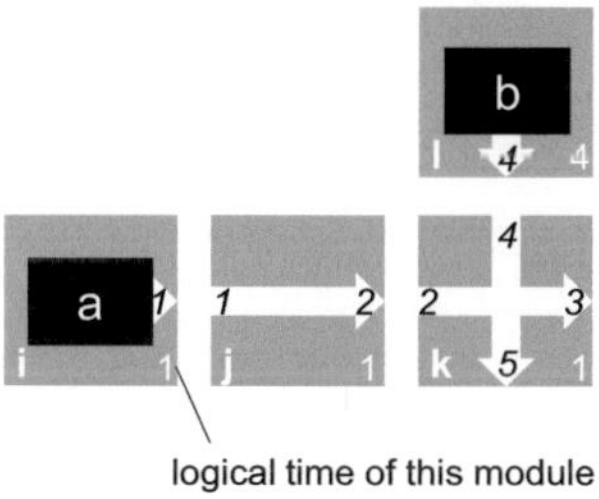

Rsv table of module k

Box	$[T_{in}, T_{out}]$	Tsp state
a	[2 , 3]	-
b	[4 , 5]	Receiving requested

Figure 4.20: Exemplary situation for granting conditions that are not fulfilled

[3] To minimize the time between two transports, it is possible to hand over monitoring of physical feasibility to the hardware control. In this case, the GRANT message would be sent before the previous box has left the module.

When are the conditions checked?

Since the conditions depend on the status of the reservation table, it is necessary to check granting conditions once there is a relevant update in the reservation table. The updates can be triggered by events related to the transport of boxes or to running reservation processes.

Relevant updates due to the transport of boxes are:

- Reception of REQUEST message: A reservation entry is newly requested to be transported.
- Reception of GRANT message for an outgoing transport: The incoming transport of another box may be performed as tandem transport.
- An outgoing transport has been finished: An incoming transport may be performed as single transport.

The relevant updates related to running reservation processes are:

- A reservation process has been successfully finished at the source module: The transport manager of the source module should send a REQUEST message for the outgoing transport to the first module on the grid.
- If the first reservation entry must be rejected because no outgoing port can be found, it is deleted from the reservation table. If the formerly second and now first reservation entry is in state RECEIVING REQUESTED, it may be performed as single transport. Figure 4.20 can serve as an example: If the reservation of box a is deleted, the incoming transport of box b can be granted.

4.4 Conclusion on Decentralized Control

The control architecture is divided into multiple components, each responsible for managing a different and distinct process: In this thesis, the reservation manager and the transport manager use the principle of logical time described in Chapter 3. Using logical time, the time-dependent route reservation leads to a more complex search tree than time-independent route reservation. For the path search, IDA* has been adapted to run on decentralized controls; we call this algorithm DIDA*.

The transport process is handled by the transport manager, whose responsibility is to adhere to the sequence of reservations defined by the logical timestamps.

System liveness is given if deadlock, livelock and starvation can be excluded. In section 3.6, we proved the absence of deadlocks. The absence of livelocks is given because we use route reservation before starting the transport of a box. The absence of starvation is closely related to the completeness of the route reservation process which has been discussed in section 4.2.6. In addition to the discussion, the observation of simulation experiments substantiate our claim of system liveness.

This chapter shows that it is possible to solve quite complex tasks such as the route planning and execution for multiple mobile objects by defining local decision rules for conveyor modules. The prevention of deadlocks with the principle of logical time has successfully been transferred to local decision rules for each conveyor module.

5 Modeling GridSorter with Agent-Based Simulation

Reality can be beaten with enough imagination.
Mark Twain

To study system behavior, agent-based simulation is used because the complex emerging behavior of several conveyor modules taking independent decisions cannot be described analytically. The simulation model is implemented with the simulation software Anylogic. Each conveyor module is represented by a software agent.

The level of abstraction from the physical system should be chosen to ensure that, on the one hand, results of the simulation are reliable and that, on the other hand, the effort required to run a simulation experiment is acceptable. Therefore, in section 5.1 the simplifications of the simulation model compared to the physical system are described.

System behavior can be observed in the simulation software in two different ways: Visualization and Parameter Variation Experiment (see section 5.2). System behavior is influenced by different input parameters which are introduced in section 5.3. Finally, system behavior is assessed with different output measures as performance indicators which are introduced in section 5.4.

5.1 Simplifications

In section 5.1.1, we present the simplifications that need to be made because multiple agents are simulated by one processor. Each agent owns its own control consisting of several control components. The principle of logical time is only applied in the reservation and transport manager. Therefore, other control components are simplified.

The general system set-up of a GridSorter system in simulation includes layout requirements and the modeling of the arriving boxes. The arrival of boxes is defined by when and where a box arrives and what destination is assigned. These simplifications are described in section 5.1.2.

By simplifying the modeling of the physical movement of boxes, the simulation effort is by far reduced because no physics engine is necessary to calculate rigid body dynamics (see section 5.1.3).

Sending of a message through several control layers on the hardware is a complex task. In the simulation, sending a message is modeled by an event. The resulting simplifications are described in section 5.1.4.

5.1.1 Simulating Decentralized Control with One Processor

To simulate the decentralized control of GridSorter, each conveyor module is modeled as an independent agent in the simulation. The agents are activated by events, for example the reception of a message or the arrival of a box. If several events happen at the same simulation time, they are activated in the order of their scheduling i.e. the order of computing of the central simulation controller.

As described in Chapter 4, the control of each module consists of several control components:

The **hardware control** is modeled without simulating sensors and actuators. The control logic is transferred to an abstract level leading to a similar behavior when transporting a box as on real hardware. The resulting simplifications are described in detail in section 5.1.3. The communication ports between modules are not modeled in the simulation. Thus, the main function of the simulation is responsible for the assignment of neighboring modules to agents.

The **layout manager** is partly included in the agent control of a conveying module. The generation of the adjacency matrix of the layout is centralized whereas the routing table is generated based on this information by each module. This hybrid approach has been chosen since topology recognition is not part of this thesis. Different algorithms have already been tested and implemented on hardware by Mayer (2009) and flexlog GmbH (2015).

As important parts of implementing the principle of logical time, the **reservation manager** and the **transport manager** are implemented as decentralized control components as described in Chapter 4.

5.1.2 General System Set-Up

The layout for a simulation experiment must fulfill the following requirements to be a valid layout: Each destination must be reachable from each source via at least one route. This means that the layout can be represented by one connected graph. If the graph is not connected, it can be divided in two or more graphs to be valid layouts again. Source modules must not be connected to more than one other module i.e. they are located outside the grid. Destination modules are allowed to have between one and four connected ports i.e. they can be located inside of the grid. In such cases, the boxes would have to be removed vertically from their destinations because they are surrounded by grid modules. One technical solution could be that the boxes fall onto some kind of slide or net.

The arrival of boxes on sources is modeled as a random process; all source modules are equally used. Each source features an infinitely large buffer for waiting boxes. Destinations are assigned to boxes randomly. Each destination is chosen with the same probability.

In the simulation, conveyor modules do not break down. This means that they are always capable of communicating with neighboring modules, taking decisions and transporting boxes.

5.1.3 Transport of Boxes

In the simulation, the transport of boxes is modeled without malfunctions. It is simplified by neglecting acceleration: The transport of a box from one module to the next has a certain duration that depends on the conveying direction. Transports in an EAST-WEST direction can have a different duration than transports NORTH-SOUTH. Additional time is needed to change the conveying direction. On the hardware, this corresponds to the change between roller and belt drive. When performing tandem transport, the middle module includes a small delay before accepting the incoming transport so that the two boxes are conveyed with a small gap between them. This corresponds to the behavior of conveying hardware. All these durations can be modeled deterministically or stochastically.

Modeling the transport of boxes with these described durations overestimates the time needed to transport boxes through several modules. In reality, the transport through several modules is faster because the boxes are not stopped and accelerated again. Modeling the real physical movements of the boxes would increase the computing time of all simulation experiments. Therefore, the transport through several modules is modeled as several single transport steps in most simulation experiments. In section 6.6, a simulation experiment is presented where the transport of boxes is modeled with acceleration and deceleration.

5.1.4 Message Sending

In the simulation, the sending of messages is also modeled without malfunctions: If a modules sends several messages to another module, the messages are received in the order they have been sent. In addition, messages are not lost between modules. A duration is assigned to the sending of a message deterministically.

5.2 Observing System Behavior

It is possible to run simulation experiments with visualization in order to observe system behavior. Figure 6.3 shows the graphical representation of the simulation environment. The visualization of system behavior together with detailed printing of information of reservation and transport process has been used for verification of the simulation model.

A parameter variation experiment can be conducted in order to analyze the system behavior under different settings. In the following section, we introduce all input parameters related to different topics.

5.3 Input Parameters

Independent variables, called input parameters in simulation, influence system behavior and are related to the general system set-up, the transport of boxes, the control algorithm and to the simulation experiment settings.

5.3.1 General System Set-Up

In order to study system behavior in different layouts (section 6.4), the **layout** for a simulation experiment can be read from an external file. Different layout criteria can have an impact on system behavior. Some measurable layout criteria are the number of modules n_{grid}, sources and destinations or the width, height and ratio of the layout. Other criteria such as the form of the layout and the position of source and destination modules are described by categories because they are not measurable quantitatively. The category N-SEW for example describes layouts with sources in the NORTH and destinations on the other three sides. A list of all layouts and their criteria can be found in Appendix B.

In order to study the system under different system loads (section 6.2.1), the **arrival of boxes** has to be modeled: Either the simulation is controlled with a constant work in process (CONWIP) i.e. there is a constant number of boxes in the simulation, or the simulation is controlled by the total arrival rate of boxes on all source modules. In this case, the inter-arrival time of boxes is modeled with an exponential distribution. In both cases, source modules dispose of infinite buffers for waiting boxes.

5.3.2 Box Transport and Message Sending

In order to model different mechanical variations of GridSorter, the physical transport of boxes can be parametrized with the following parameters: the duration of a transport step in East-West direction, the duration of a transport step in North-South direction, the duration of a direction change and the time delay between two boxes if conveyed in tandem. In reality, acceleration and deceleration influence transport times. Consequently, the transport from one module to the next one takes shorter if the box does not need to stop but can move through several modules. In a second modeling of box transport, transport times vary dependent how far a box is transported without being stopped. In addition, the transport times of two boxes are adjusted if they are transported in tandem.

The sending of messages can be parametrized by the duration of the message transfer to a neighboring module. To exclude the impact of communication effort on system performance the duration of message transfer can be set close to zero.

5.3.3 Parameters for Basic Control Algorithm

In Chapter 4, different possibilities have been presented for variation of the basic control algorithm. To study their impact on system performance, the following parameters need to be set. The study is presented in section 6.2.2.

One main characteristic of the box transport is whether several boxes can move together in the same direction (cf. section 3.2). If the hardware control is able to perform tandem transport, the **tandem transport settings** can be set as follows:

- **S** Tandem transport is not allowed; only single transport is possible.
- **T** Tandem transport is allowed and used if possible.

The **limitation of the path length** aims to reduce the size of the search tree by limiting routing possibilities (cf. page 72). It can be set as follows:

- **Pa** The path length is not limited. Each module is allowed to forward a reservation request to all connected ports.
- **Pb** The path length is limited to the minimal path length between source and destination with the following exception: If there is only one direction promising the minimal path length to destination, the directions with the second minimal path length can be chosen as well.
- **Pc** The path length is limited to the minimal path length[1] between source and destination: Each module is only allowed to forward a reservation request to the ports with the minimal path length to destination.

The reachability of modules depending on this setting is shown in Figure 5.1 for two exemplary layouts.

The **limitation of routes through source neighbors** aims to hinder boxes from blocking sources (cf. page 73). It can be set as follows:

- **Na** The routes through source neighbors are not limited. Each connected module is allowed to send reservation requests to source neighbors.
- **Nb** Connected modules are only allowed to send reservation requests to source neighbors if they are a source or source neighbor, or if this direction is the only direction offering the minimal path length to destination.

[1] The path length is measured in transport steps, i.e. the number of direction changes may be different.

- **Nc** Connected modules are only allowed to send reservation requests to source neighbors if they are source or source neighbor.

Exemplary routes depending on this setting are shown in Figure 5.2.

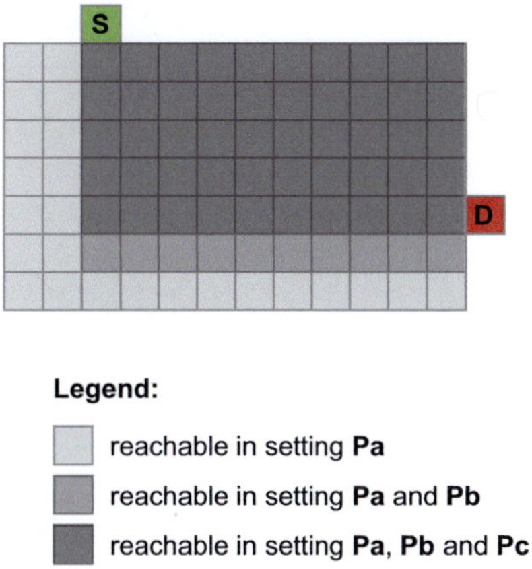

Figure 5.1: Reachable modules in different settings for limitation of path length (letters correspond to description of settings)

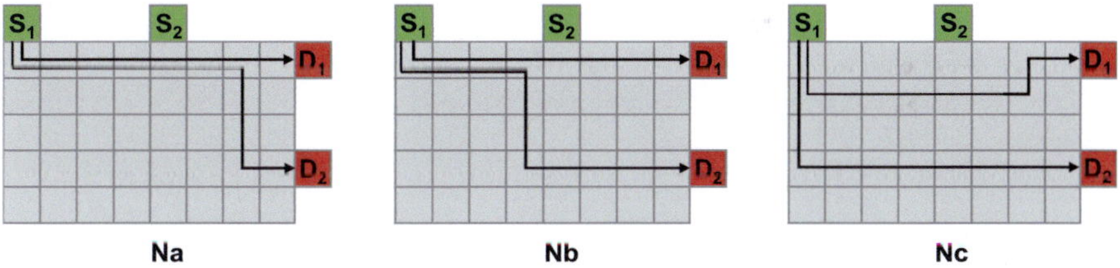

Figure 5.2: Exemplary routes in different settings for routes through source neighbors (letters correspond to description of settings)

The **criterion for selecting the outgoing port of source neighbors** also aims to hinder boxes from blocking sources (cf. page 74). It can be set as follows:

- **lt** Source neighbors select the port with the lowest expected lead time (like all other modules).
- **ts** Source neighbors select the port with the lowest timestamp for outgoing transport.

5.3.4 Parameters for Partial Route Reservation

For the extension of the algorithm to allow partial route reservations as described in section 4.2.5, two parameters are relevant: If a source module needs to start a new search iteration by increasing the maximal lead time, it can decide whether an inter-destination should be searched for. If the number of messages that have been already sent within this reservation process exceeds the chosen **message count limit**, the search for an inter-destination is started. One criterion to select an inter-destination is the **maximal remaining distance to destination**. The impact of these parameters is studied in Appendix C. In section 6.5, the impact of partial route reservation on system performance and communication effort is examined.

5.3.5 Simulation Experiment

Each simulation run is parametrized by its **random seed** which is used for the random number generators. The random number generators are only used to model the box arrival and the destination assignment. The control algorithm is purely deterministic.

The simulation time is measured in seconds. The **duration of one simulation run** and the **number of replications** can be parametrized. To guarantee steady-state characteristics, all output variables are only recorded after the **duration of the warm-up period**. The input parameter related to the simulation experiment are determined and statistically reasoned in section 6.1.1.

5.4 Performance Indicators

Different dependent variables are recorded in order to calculate performance indicators and to make system behavior interpretable. The dependent variables are either related to the total system, or to each single module or to each single box. All dependent variables are determined by Monte Carlo simulation since they cannot be analytically calculated. Unless otherwise stated, average values are calculated during one simulation run for all measures. For these to be reliable, the system must be stable i.e. not overloaded.

For the application of the system in real world scenarios, mainly the system throughput is important. To assess the efficiency of the control algorithm, the number of sent messages is important because it gives an insight into the duration of a reservation process and the workload of one control. In addition to these measures, we also record other measures such as the occupancy of modules to allow interpretation of system behavior based on the simulation results. Logical clocks are only synchronized to each other when performing a transport step together. Consequently, the set of logical clocks is asynchronous from the point of view of physical time. In order to better understand system behavior, several measures are recorded related to synchronicity of logical clocks.

We describe below all recorded measures depending on the object they are related to.

Related to the total system, the following variables are recorded to assess system performance:

- Throughput λ_{tot}: number of boxes reaching their destination per time interval.
- Number of boxes in the system N_{tot}: number of boxes on the grid or waiting on source modules or their infinite buffer. If the arrival of boxes is controlled by CONWIP, N_{tot} is an input parameter.
- Number of boxes on the grid N_{grid}: boxes located on the grid including the source and destination modules.
- Number of waiting boxes N_{wait}: boxes waiting for source modules to get available (per source or in total). This variable indicates whether the system is overloaded.

Related to a group of modules, the following variables are recorded to assess synchronicity of logical clocks:

- Mean logical time $\bar{C}_g(t)$ of all modules of group g at simulation point of time t
- Average absolute deviation of logical time of one module from mean logical time of group g of modules

$$d_{abs,g} = \frac{1}{n_g}\frac{1}{T_{sim}}\sum_{t=0}^{T_{sim}}\sum_{n_g}|\bar{C}_g(t) - C_i(t)|$$

- Average difference between mean logical time of group g and mean logical time of group h

$$d_{g,h} = \frac{1}{T_{sim}}\sum_{t=0}^{T_{sim}}(C_h(t) - C_g(t))$$

Related to one module, the following variables are recorded to better visualize system behavior:

- Occupancy o_i: Percentage of time the conveying module i is occupied by a box.
- Relative throughput λ_i: Number of boxes transported by module i per time interval as ratio of system throughput per time interval.

Related to one box, we record the following variables to assess communication effort and system performance:

- Number of messages per reservation m_{rsv}
- Number of messages per reservation $m_{rsv,1}$ that are sent before the first transport starts. If the complete route is reserved before the transport process, it holds $m_{rsv,1} = m_{rsv}$
- Real lead time of the performed transport process from source to destination measured in physical time t_{real}
- Difference between expected lead time measured in logical time and real lead time measured in physical time t_diff
- Path length from source to destination measured in transport steps l_{tot}

Interrelation between different indicators The system throughput can be calculated by the sum of the throughput of all sources or the sum of the throughput of all destinations (flow equality):

$$\lambda_{tot} = \sum_{src} \lambda_i = \sum_{dest} \lambda_i$$

The total number of boxes in the system corresponds to the sum of boxes on the grid and waiting boxes

$$N_{tot} = N_{grid} + N_{wait}$$

Little's Law describes the interdependence between the number of boxes in the system, the throughput of the system and the lead time of one box. With the described indicators, Little's law can be expressed by

$$\lambda_{tot} = N_{grid}/t_{tot} \,.$$

The sum of the occupancy of all modules of the system equals the number of boxes in the system

$$\sum_{grid} o_i = N_{grid} \, .$$

The described measures are used in Chapter 6 to assess the system behavior when answering different research questions.

6 System Behavior of GridSorter Controlled with Logical Time

After climbing a great hill, one only finds
that there are many more hills to climb.
-- Nelson Mandela

In this chapter, we study system behavior of GridSorter under different settings in order to understand the influence of different input parameters on system performance. In a first step, we present the general simulation set-up, the statistical analysis of results and the approach of selecting layouts for different simulation studies. In section 6.2, we first illustrate the basic system behavior with different plots and charts in order to make it better interpretable. We then determine the combination of basic control parameters that leads to the best system performance. In section 6.3, the synchronicity of logical clocks and its influence on system performance is studied. In section 6.4, we investigate the influence of layout characteristics on system performance and communication effort. Whether partial route reservation reduces communication effort is studied in section 6.5. In section 6.6, we study system behavior under varying transport times that arise from acceleration and deceleration during transport of boxes.

6.1 General Simulation Set-Up

In this chapter, we use the following terms to describe our simulation approach: A *simulation run* represents the uninterrupted simulation of the system during a certain time period. Each simulation run is replicated with different random seeds but identical settings. The term *simulation experiment* includes these replications. The results of one simulation experiment are determined by calculating mean values of the replicated simulation runs. In a *simulation study*, multiple simulation experiments under different settings are conducted so that the influence of the relevant input parameters on system behavior can be investigated.

In all following simulation experiments, the settings for the arrival of boxes and the experiment settings are identical. The choice of these parameters is reasoned in the next section.

- The arrival of boxes is controlled with CONWIP which guarantees a stable system.
- The performance indicators are only recorded after a warm-up phase of 200 seconds.
- The results of one simulation run correspond to one hour of operation in physical time i.e. 3600 seconds.
- Each simulation run is replicated 10 times with different random seeds. These runs together are one simulation experiment.

6.1.1 Statistical Analysis of Simulation Results

For the statistical analysis, two questions have to be considered:

1. Is the combination of duration of one simulation run and number of iterations sufficient to get **reliable mean values**?
2. Is the warm-up period long enough to observe the system exclusively in **steady-state**?

Let us treat the first question: To measure the **reliability of results**, the standard error of the estimated mean is calculated for each simulation experiment. The relative standard error of the mean is defined by:

$$\mathrm{SE}_x = \frac{s}{\overline{x}\sqrt{N}} \tag{6.1}$$

where s is the sample standard deviation, $\overline{x}$ the sample mean and N number of samples.
The sample standard deviation is defined by:

$$s = \sqrt{\frac{1}{N-1}\sum_{i=1}^{N}(x_i - \overline{x})^2} \tag{6.2}$$

To determine reliability of our results, we use system throughput as most important performance indicator.

Figure 6.1 represents the frequency of values of the relative standard error of system throughput for all 4663 simulation experiments of the following sections of this thesis. The relative standard error is less than 1.4% for all simulation experiments and for most experiments, it lies between 0.2% and 0.4%. We deduce that 10 simulation experiments are sufficient to determine the mean system throughput of a simulation experiment.

Let us now treat the second question: The warm-up period should guarantee that the system is in **steady-state** when the recording of the performance indicators is started. In our system, steady-state is achieved if the number of boxes on the grid corresponds to the average number of boxes on the grid in steady-state. The number of boxes on the grid has been chosen as indicator for steady-state because all experiments are controlled with CONWIP.

In layouts with a high number of modules, it takes longer to achieve this system state. When observing the layout with the highest number of modules, one can see that the number of boxes in the system corresponds to the average number of boxes in the system after 25 seconds already. We therefore assume that a warm-up period of 200 seconds is sufficient to guarantee steady-state results at the start of recording.

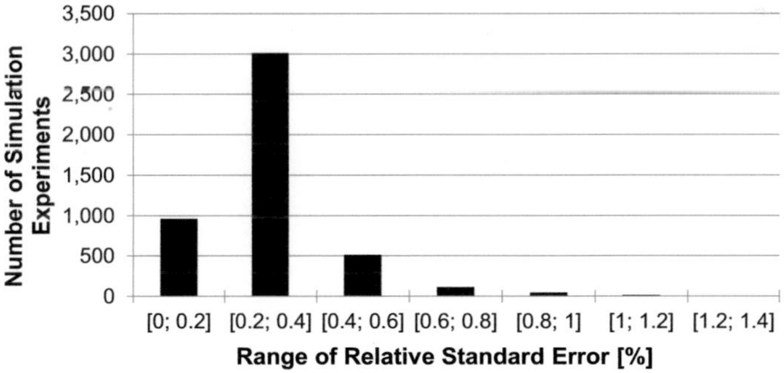

Figure 6.1: Relative standard error of all simulation experiments

6.1.2 Selecting Layouts for Simulation Studies

How are the layouts selected for different simulation studies? Layouts can be described by many different characteristics such as number of modules, number of sources and destinations, shape of the layout, or positioning of sources and destinations. In addition, not only absolute numbers are relevant, but the ratio of different numbers can be relevant as well. An example: It is not suitable to compare a layout with 25 modules and 6 sources with a layout of 500 modules and 6 sources to investigate the influence of layout size, because in the big layout, sources would represent the bottleneck. In this case, it is more suitable to maintain the ratio of number of sources and number of modules.

For the simulation studies, a layout population of 85 layouts has been generated. Since we focus on dense networks, the shape is always a rectangle or close to one (some layouts have "cropped" corners). Likewise, in most layouts the number of destinations exceeds the number of sources which is a typical situation for sorting of goods. In most layouts, sources are positioned on the NORTH border of layout and destinations on the remaining three borders. In the layout presented in Figure 6.2, the sources and destinations are positioned in this way which is called N-SEW.

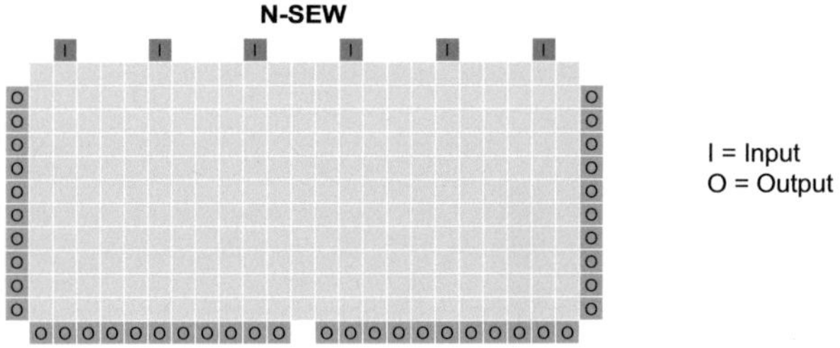

Figure 6.2: Graphical representation of one layout

In few layouts, the positioning of sources and destinations is different. All other characteristics are varied so that the population of layouts is as heterogeneous as possible. The number of modules varies from 25 to 506, the number of sources from 1 to 36 and the number of destinations from 15 to 65. The aspect ratio of the layouts varies from 0.8 to 8. The population of layouts can be studied in more detail in Appendix B where all layouts are graphically presented and where the table in Tables B.2, B.3 and B.4 shows an overview of their characteristics.

For some simulation studies, additional layouts have been created in order to show specific effects. Those layouts are introduced in the relevant sections.

6.2 System Behavior and Basic Control Settings

In this section, we study different aspects of basic system behavior. To exclude the influence of varying transport times, the durations related to box transport are chosen in the most simple way: Regardless of direction, each transport step takes 1 second and each direction change takes 0.2 seconds. Time delay between two boxes conveyed in tandem is 0.05 seconds. The message transfer time is set to 1^{-7} milliseconds which actually corresponds to immediate sending of messages. Like this, system behavior is not negatively influenced by the duration of reservation process. In section 6.5, we then set the message transfer duration to the more realistic value of 3 milliseconds in order to show the effectiveness of partial route reservation.

6.2.1 Description of System Behavior

The objective of this section is to illustrate system behavior in more detail. This is why we run experiments with the layout shown in Figure 6.2 and vary the total number of boxes in the system. In this way, we can see how system behavior changes dependent on system load. The basic input parameters are set to the combination **T Pb Nb ts** (described in section 5.3.3). The choice of this combination is reasoned in section 6.2.2 with a simulation experiment using full factorial design and the big layout population.

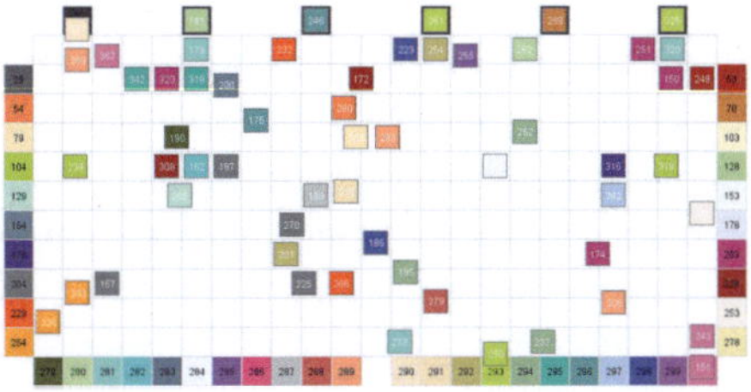

Figure 6.3: Simulation screen shot of the layout of Figure 6.2

Figure 6.3 shows a screen shot of simulation of the layout presented in Figure 6.2 under high system load. All sources are occupied, and boxes waiting for sources to become available are not shown. The number on the boxes depicts the ID of the destination conveyor. Since a screen shot does not help much to understand system behavior, we present different diagrams in the following.

Performance Indicators Related to the Total System

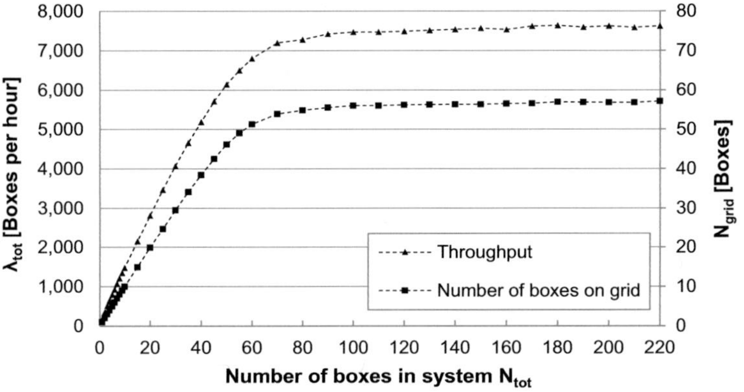

Figure 6.4: Throughput and number of boxes in system under different system load

Figure 6.4 shows throughput and number of boxes on the grid in relation to the total number of boxes in the system i.e. the system load (CONWIP). We can see that the number of boxes on the grid first increases linearly with system load because all boxes can be transported. Beginning from 50 boxes as system load, some boxes have to wait before being transported. The system is then saturated with around 57 boxes on the grid; between every fourth and fifth module is occupied with a box.

Throughput shows a very typical trend: It first increases strongly with system load and then approaches its limit around 7600 boxes per hour. It is important to notice that overloading the system does not have a negative impact on system performance. Therefore, no regulation of load is needed to maximize throughput. This effect has also been observed for other layouts. In the simulation studies of following sections, system load is chosen very high so that maximal throughput is achieved.

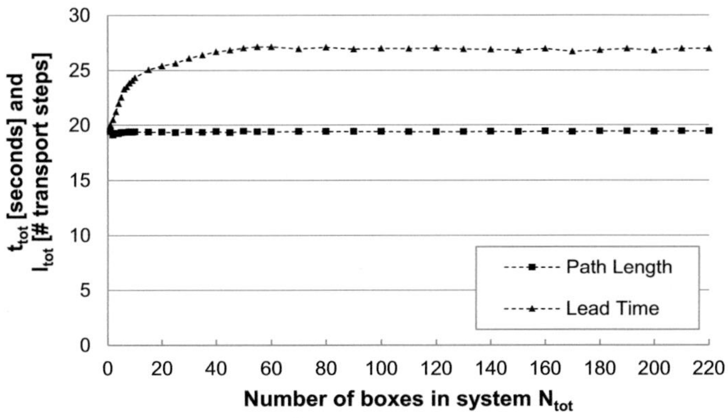

Figure 6.5: Lead time and path length under different system load

Figure 6.5 shows average lead time and average path length in relation to the total number of boxes in system. If one box is in the system, lead time and path length are almost identical; the difference originates from the duration of direction changes which is not included in path length. Lead time increases with system load whereas path length remains at around 19.4 transport steps. The stability of path length can be explained by the input parameter setting **Pb** limiting path length to minimal and second minimal path lengths. The increase of lead time originates solely from an increase in waiting time. The strong increase of lead time under low system loads will be explained by the synchronicity of logical clocks in section 6.3.

Performance Indicators Related to the Modules

Figure 6.6 illustrates occupancy and relative throughput of two different layouts (left and right) under two different system loads. Heat maps have been chosen as illustration method: The gray value of one module represents its occupancy o_i or its relative throughput λ_i. What can be observed from these plots?

- The left layout is already used to its full capacity with the low system load of 20 boxes because the two sources are the bottleneck. Consequently, the occupancy plot under high system load of 220 boxes is quite similar.

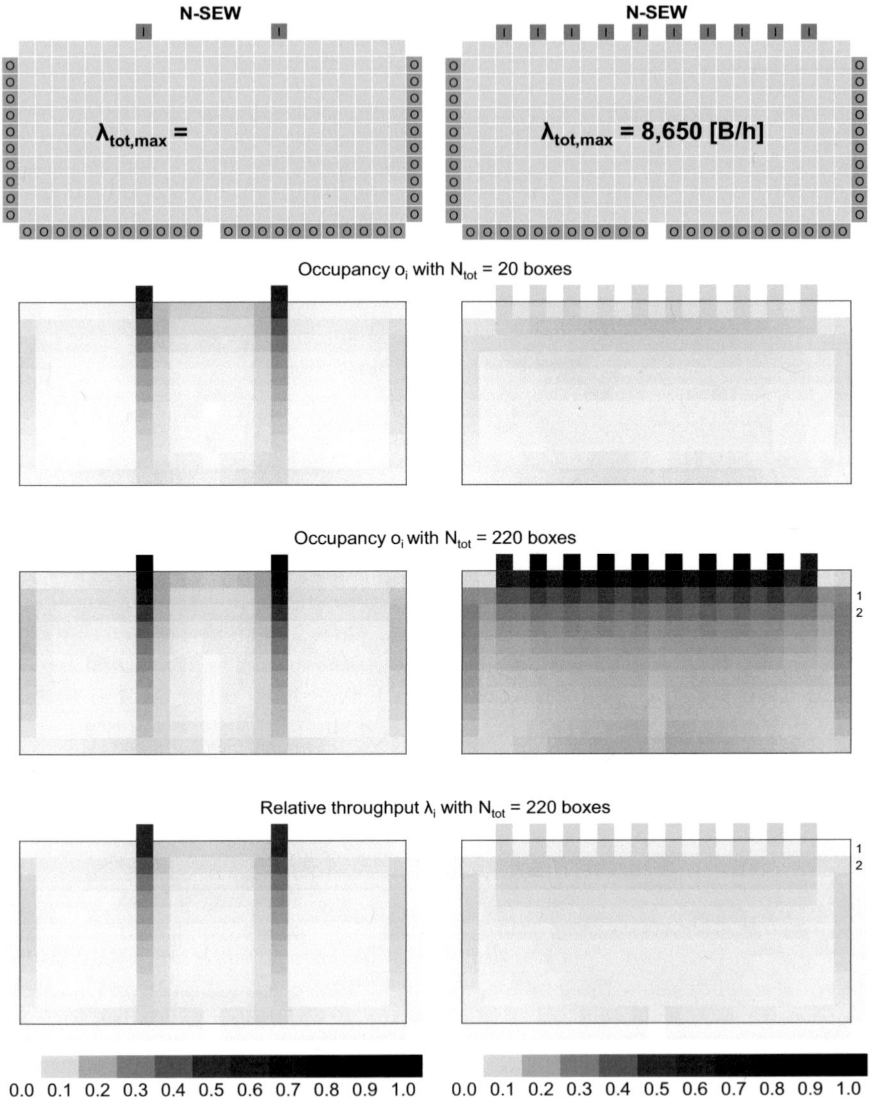

Figure 6.6: Heat maps for occupancy and throughput of different layouts under different system loads

- In contrast, the right layout is able to receive more boxes thanks to 10 sources: the occupancy plot under high system load is much darker than the occupancy plot under low system load.
- In general, occupancy is higher around source modules which indicates that they are the bottleneck of the layouts. This seems logical since there are much more destinations than sources in both layouts.
- The last row illustrates relative throughput of both layouts under high system load[1]. Again, it is evident that the left layout is not uniformly used whereas the right layout shows a very homogeneous usage of the modules.
- The throughput plots of both layouts show that source modules are used equally.
- When comparing the occupancy and throughput plots under high system load, they look quite similar: With high throughput, a module is often occupied. However, a module could also be occupied by waiting boxes and therefore achieve low throughput. This effect can be observed within the second row of the grid of the right layout: Throughput is higher than in the first row but occupancy is lower. This means that boxes wait on source modules and on modules of the first row for the second row to become available. In the second row, additional throughput is achieved by EAST-WEST-movement of boxes.

Figure 6.7 illustrates occupancy under high system load for two layouts with another positioning of sources and destinations. What can be observed?

- In Figure 6.6 and 6.7, the modules that are aligned with source modules are more occupied than other modules. This can be explained by the routing preference to use the opposite port of the incoming port as outgoing port to avoid direction changes.
- The occupancy plot of the right layout is generally darker which means that there are more boxes on the layout which leads to a higher throughput. In general, one can say that a layout has a good structure and shape, if all modules are used uniformly i.e. if the occupancy and throughput plots show homogeneous gray values. The darker the occupancy plot in total, the more boxes are on the layout.

[1] Since the throughput plot under low system load looks quite similar, it has not been included.

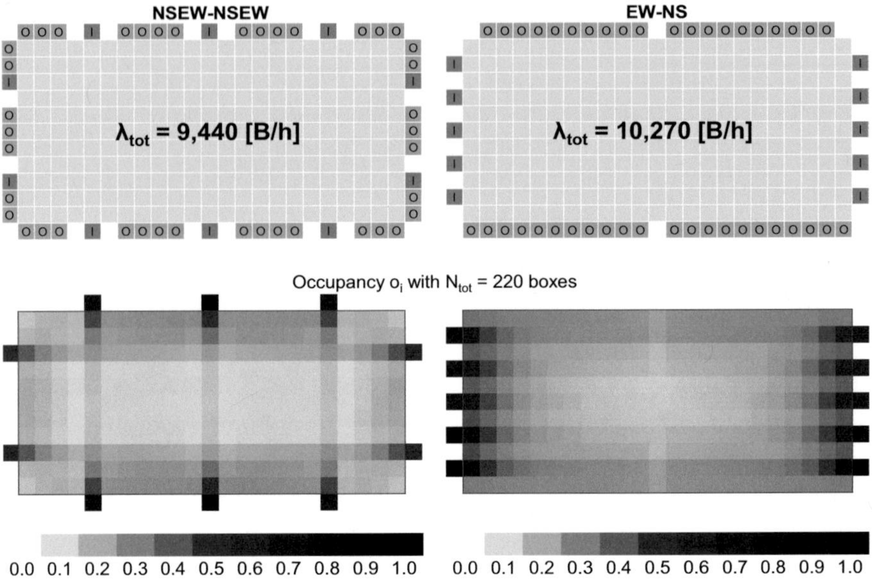

Figure 6.7: Heat maps for occupancy of layouts with scattered sources and destinations

6.2.2 Setting of Basic Control Parameters

The objective of this section is to find a combination of settings for basic control parameters that leads to high throughput with low communication effort regardless of the layout. The relevant input parameters are described in section 5.3.3 and result in 36 possible combinations. Simulation experiments have been conducted with each parameter combination for all 85 layouts of our layout population. The number of boxes in the system is set so high that there are always boxes in front of source modules. The simulation experiments have been conducted together with Freund (2015).

We compare the parameter combinations in different ways. First of all, we count the number of layouts for which a certain parameter combination leads to the maximal throughput in this layout. This measure indicates in how many layouts the parameter combination leads to best system performance. Figure 6.8 illustrates that the combination **T Pb Nb ts** achieves maximal throughput in 20 out of 85 layouts.

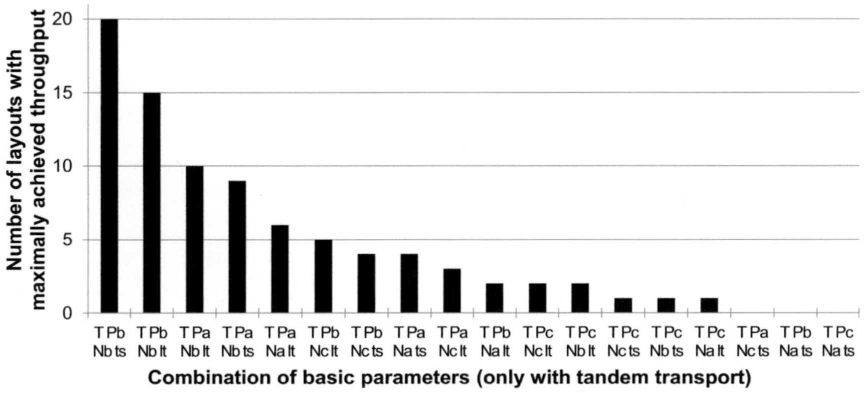

Figure 6.8: Number of layouts with maximally achieved throughput for each parameter combination (see section 5.3.3 for notation of settings)

Second, we determine for each layout the maximally achieved throughput and express the throughput achieved by any parameter combination as proportion of this maximal throughput. Like this, the performance of a parameter combination in all layouts is made comparable despite the difference of layouts. Likewise, the communication effort is made comparable: We calculate the proportion of number of messages per reservation compared to the minimally achieved value in this layout. With these measures, an analysis of variance (ANOVA) is conducted.

Figure 6.9 shows the results of the analysis of variance for different input parameters. In the first table, for example, all parameter combinations with tandem transport are compared to all parameter combinations without tandem movement. The left table analyzes the throughput ratio whereas the right table analyzes the message ratio. The best performing group is highlighted in gray in each table. A low p-value indicates that the null hypothesis can be rejected with a low significance level concluding that there is strong evidence that the groups differ from each other.

The following insights can be read from Figure 6.9:

- The system achieves higher throughput if tandem transport is allowed in setting **T**.
- The system achieves lower throughput with strict limitation of path length in setting **Pc**. The difference of throughput ratio between **Pa**

and **Pb** is not significant but communication effort with setting **Pb** is significantly lower.

- Both settings **S** and **Pc** reduce the size of the search tree. On the one hand, this reduces communication effort but, on the other hand, it also reduces throughput.
- The system achieves higher throughput if routes through source neighbors are slightly restricted in setting **Nb**.
- The difference of throughput ratio among the settings **ts** and **lt** is not significant. Both criteria for selecting the outgoing port of source neighbor modules lead to comparable results.

Throughput as ratio of maximally achieved throughput

Transport Settings

	μ	σ^2	F	p-value
S	87.75%	0.30%	1986.27	0.000
T	96.36%	0.27%		

Path Length Limit

	μ	σ^2	F	p-value
Pa	93.17%	0.30%	100.64	0.000
Pb	93.35%	0.32%		
Pc	89.65%	0.70%		

	μ	σ^2	F	p-value
Pa	93.17%	0.30%	0.54	0.462
Pb	93.35%	0.32%		

Routes through Source Neighbors

	μ	σ^2	F	p-value
Na	90.76%	0.48%	31.87	0.000
Nb	93.12%	0.50%		
Nc	92.28%	0.41%		

	μ	σ^2	F	p-value
Nb	93.12%	0.50%	7.97	0.005
Nc	92.28%	0.41%		

Criterion at Source Neighbors

	μ	σ^2	F	p-value
ts	92.01%	0.47%	0.15	0.697
lt	92.10%	0.47%		

Number of messages as ratio of minimally achieved number of messages

Transport Settings

	μ	σ^2	F	p-value
S	124.87%	10.15%	65.06	0.000
T	133.98%	9.37%		

Path Length Limit

	μ	σ^2	F	p-value
Pa	147.85%	20.07%	327.92	0.000
Pb	123.25%	2.56%		
Pc	117.17%	2.00%		

	μ	σ^2	F	p-value
Pa	147.85%	20.07%	272.67	0.000
Pb	123.25%	2.56%		

Routes through Source Neighbors

	μ	σ^2	F	p-value
Na	113.96%	1.25%	283.89	0.000
Nb	129.77%	5.01%		
Nc	144.55%	18.97%		

	μ	σ^2	F	p-value
Nb	129.77%	5.01%	92.90	0.000
Nc	144.55%	18.97%		

Criterion at Source Neighbors

	μ	σ^2	F	p-value
ts	128.83%	9.93%	1.08	0.298
lt	130.02%	10.00%		

Figure 6.9: Results of analysis of variance for basic control parameters (see section 5.3.3 for notation of settings)

These insights are also confirmed by comparing the throughput ratios of all 36 parameter combinations. The top-4 parameter combinations (combining **Pa**, **Pb** and **ts**, **lt**) are compared by an analysis of variance in Figure 6.10. The low p-value indicates that there is strong evidence that the groups differ from each other. The parameter combination **T Pb Nb ts** performs best by achieving in average 99.41% of the maximally achieved throughput. Related to communication effort, it performs second best by achieving in average 129.3% of the minimally achieved number of messages. The differences in throughput and communication effort between the first and second best parameter combination are very small.

It is interesting that both parameters that restrict routing possibilities lead to best performance with the middle setting **Pb** and **Nb** which is a compromise between no a strict limitation. Based on the results of both comparison methods, all following simulation studies are conducted with the parameter combination **T Pb Nb ts**. This combination is defined by the following setting:

- **T** Tandem transport is allowed.
- **Pb** The path length is limited to the minimal path length between source and destination with the following exception: If there is only one direction promising the minimal path length to destination, the directions with the second minimal path length can be chosen as well.
- **Nb** Connected modules are only allowed to send reservation requests to source neighbors if they are a source or source neighbor, or if this direction is the only direction offering the minimal path length to destination.
- **ts** Source neighbors select the port with the lowest timestamp for outgoing transport.

Throughput as ratio of maximally achieved throughput

Top 4 Parameter Combinations	μ	σ²
T Pb Nb ts	99.41%	0.01%
T Pb Nb lt	99.22%	0.01%
T Pa Nb lt	99.00%	0.01%
T Pa Nb ts	98.73%	0.02%

F	p-value
7.22	0.0001

Number of messages as ratio of minimally achieved number of messages

	μ	σ²
T Pb Nb ts	129.29%	1.16%
T Pb Nb lt	128.92%	1.24%
T Pa Nb lt	153.10%	6.63%
T Pa Nb ts	147.52%	4.13%

F	p-value
40.05	0.000

Figure 6.10: Results of analysis of variance for top-4 parameter combinations (see section 5.3.3 for notation of settings)

6.3 Synchronicity of Logical Clocks

In contrast to physical clocks, logical clocks in our system are only set forward if a box is transported. Consequently, logical clocks do not show identical values of logical time. If the logical time of different modules differs strongly from each other, it can lead to over- or underestimating of the expected lead time during the route reservation process. This means that the real lead time during transport process differs from the expected lead time. Since the expected lead time is measured in logical time, it has converted to physical time which is done using the duration of transport steps.

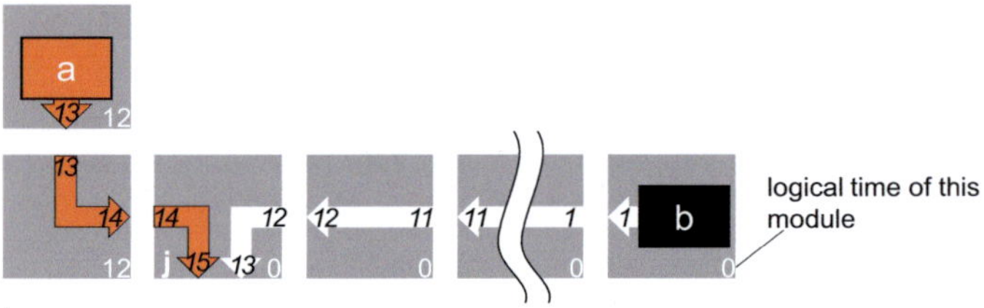

Figure 6.11: Exemplary situation for waiting because of asynchronous logical clocks

Figure 6.11 shows an exemplary situation where the expected lead time underestimates the real transport lead time. The logical clocks of the source modules of both boxes differ strongly. During reservation process, box b is planned to use module j before box a. During transport process box a will be arriving first and has to wait until box b has passed. Consequently, the real lead time of box a will be higher than its expected lead time. We can see that waiting times arise from over- or underestimating real lead time. And these waiting times can have a negative impact on throughput i.e. system performance. We therefore study synchronicity of logical clocks in this section using different measures related to different groups of modules. To examine the impact of asynchronous logical clocks, we compare the results of systems where logical clocks are only synchronized by box transports with systems where logical clocks of source modules are additionally synchronized with each other.

6.3.1 Synchronization Based on Box Transport

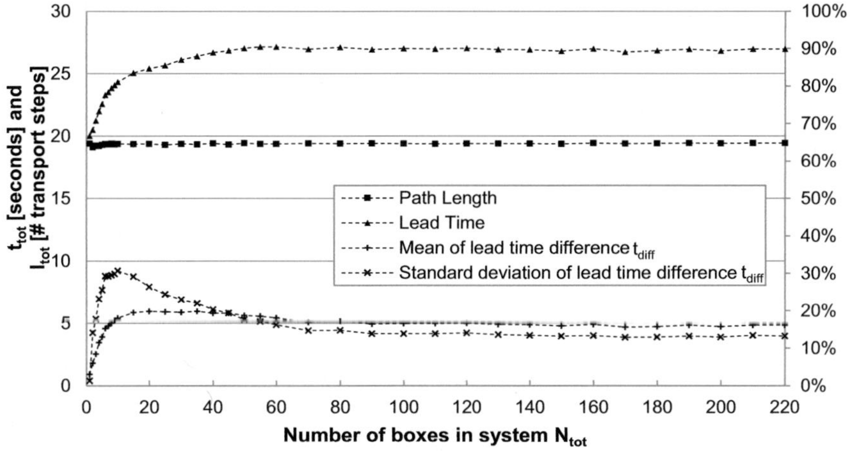

Figure 6.12: Lead time and path length and standard deviation of lead time difference under different system load

Figure 6.12 resumes Figure 6.5 by adding the mean and standard deviation of the difference between expected and real lead time t_{diff} as portion of the real lead time. Under all system loads, the expected lead time tends to underestimate the real lead time. High standard deviation indicates that the expected lead time strongly over- and underestimates real lead time. The highest standard deviation is achieved with around 10 boxes in the system. The strong increase and following decrease of deviation represents an explanation for the bend in increase of lead time. Two effects influence the lead time contrariwise with increasing system load: The routes of boxes interfere with each other which leads to "real" waiting times and an increase in lead time. However, logical clocks are also better synchronized which reduces the difference between expected and real lead time and therefore decreases lead time. The better synchronicity of logical clocks with increasing system load is shown in Figure 6.13. It visualizes the average absolute deviation of a logical clock among the logical clocks of all modules $d_{abs,all}$, among source modules $d_{abs,sources}$, among grid modules $d_{abs,grid}$ and among destination modules $d_{abs,dest}$.

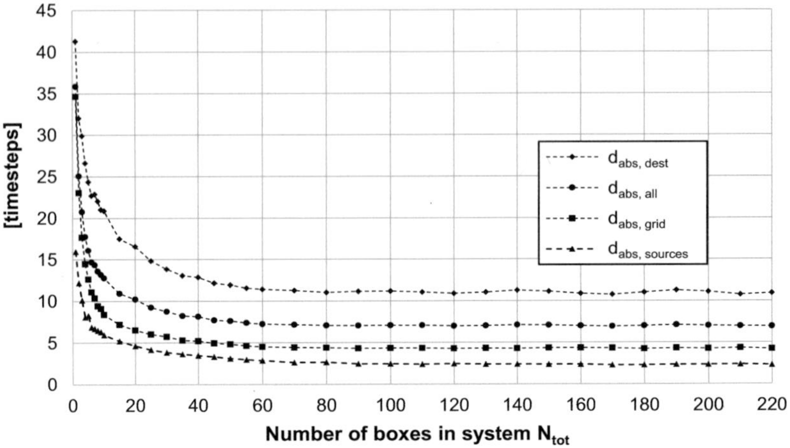

Figure 6.13: Absolute deviation of logical time of different groups of logical clocks in the standard layout

Figure 6.13 shows, as expected, that synchronicity of logical clocks increases with increasing system load because more transports are performed for which the logical clocks of participating modules need to be synchronized. We can also see that source modules are generally better synchronized with each other than the group of all modules. Under full system load, source modules deviate around 3 to 4 time steps from the mean.

Where does the strong asynchronicity under low system loads come from? In the following, we show that the behavior of logical clocks strongly depends on the positioning of sources and destinations. As example layouts, we have chosen the standard N-SEW layout of Figure 6.2 and the corresponding NSEW-NSEW layout.

Before we go into detailed explanation, we first want to present three general effects regarding synchronicity of logical clocks:

- **Neighbors** Modules with high number of adjacent modules are synchronized more often i.e. their logical clock is more often set forward. Due to this effect, grid modules tend to be ahead in logical time.
- **Throughput** Modules with high relative throughput perform transports more often i.e. their logical clock is more often set forward.

– **Distance to destination** Timestamps of transport steps must increase with each transport step. The timestamp of the transport step of one specific box to its destination is therefore higher than the timestamp of the transport step of the source module. The difference between both timestamps increases with path length.

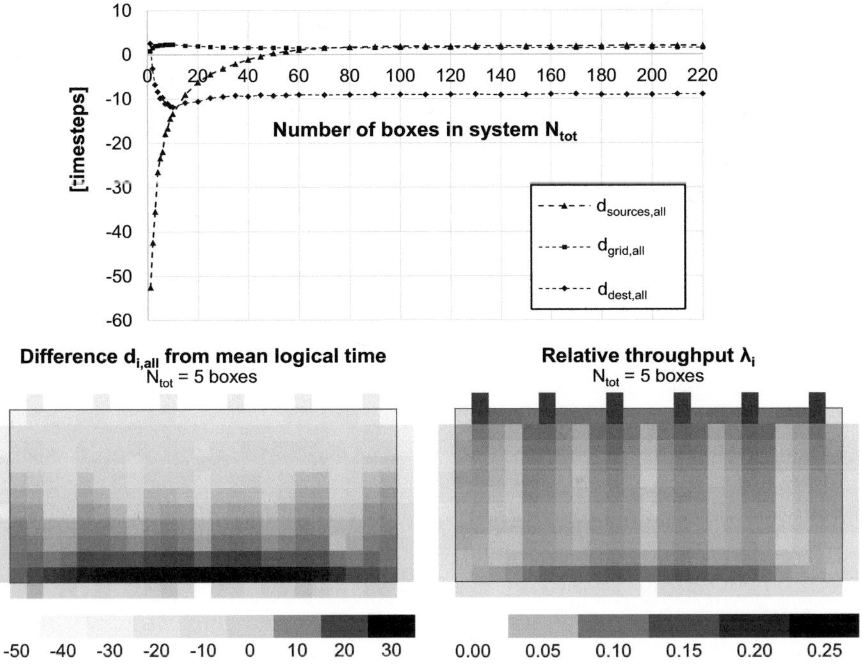

Figure 6.14: Average logical time of a N-SEW layout under different system loads (upper part) and difference of logical time of modules (left) and relative throughput (right) under low system load

Figure 6.14 shows the average difference of logical time of a group of modules from the mean of all logical clocks in the layout. The measure is recorded for sources $d_{sources,all}$, grid modules $d_{grid,all}$ and destinations $d_{dest,all}$. If the difference is positive, the clocks of this group tend to go ahead. If it is negative, the clocks of this group tend to lag behind. The heat map on the left side shows the average difference of each single logical clock from the mean logical time $d_{i,all}$ under system load of 5 boxes. The heat map on the right side shows the relative throughput of each module.

Under low system load, source modules tend to lag behind and destination modules are ahead of source modules. Under these conditions, the described effect of **distance to destination** must be outperforming the effect of the higher **throughput** of source modules. Analogous, the left heat map shows a smooth rise of logical time from NORTH to SOUTH. In the southern part of the layout, the relative **throughput** is high which leads to clocks going ahead due to the second effect. With only one box in the system, the logical clocks of destination modules are even ahead of the logical clocks of grid modules. Since there is only one box in the system, it always takes the same path with shortest path length and least direction changes to a destination. Consequently, the first effect due to synchronization with **neighbors** is weak and the second effect is weak because all the modules on this route are equally used. The third effect leads to destinations being ahead and sources lagging behind. Under high system load, destination modules lag behind because of low relative **throughput**. The logical time of source and grid modules go ahead in similar way.

Figure 6.15 shows the same plots but for the corresponding NSEW-NSEW layout. The logical clocks show different, almost contrary behavior: Under low system loads, source modules go ahead and destination modules lag behind.

Because sources and destinations are positioned close to each other, the grid modules close to source modules go ahead due to the low **distance to destination** as can be seen in the heat map of logical time. In addition, the relative throughput is high in the southern and northern part of the layout. Consequently, sources are even more ahead because the effect of relatively high **throughput** and short **distance to destination** do not compensate such as in the other layouts but sum up. For high system loads, destination modules lag behind as in the N-SEW layout, but sources go more ahead than grid modules.

Even though the behavior of logical clocks under low system loads is different depending on the positioning of sources and destinations, synchronization increases with system load in all layouts. Nevertheless, expected lead time and real lead time still differ from each other.

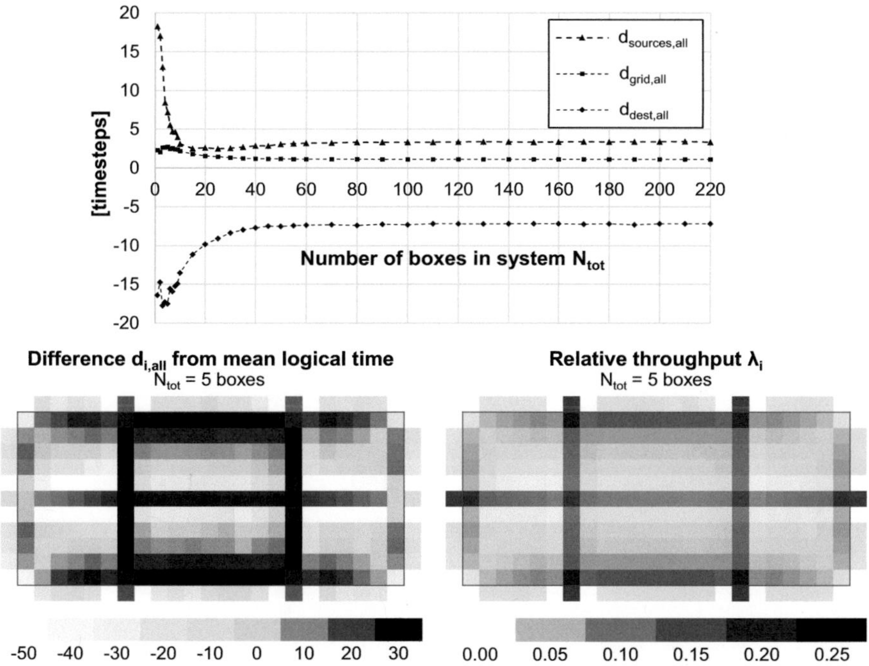

Figure 6.15: Average logical time of a NSEW-NSEW layout under different system loads (upper part) and difference of logical time of modules (left) and relative throughput (right) under low system load

6.3.2 Additional Synchronization Among Source Modules

As can be seen in the example situation of Figure 6.11, asynchronous logical clocks of sources lead to waiting times during transport. We therefore implement an additional synchronization among source modules: Each time a source module sets its logical clock forward, it informs all other source modules. Those set their logical clock forward accordingly if they have no reservations for lower timestamps. Like that, the logical clock condition is still fulfilled.

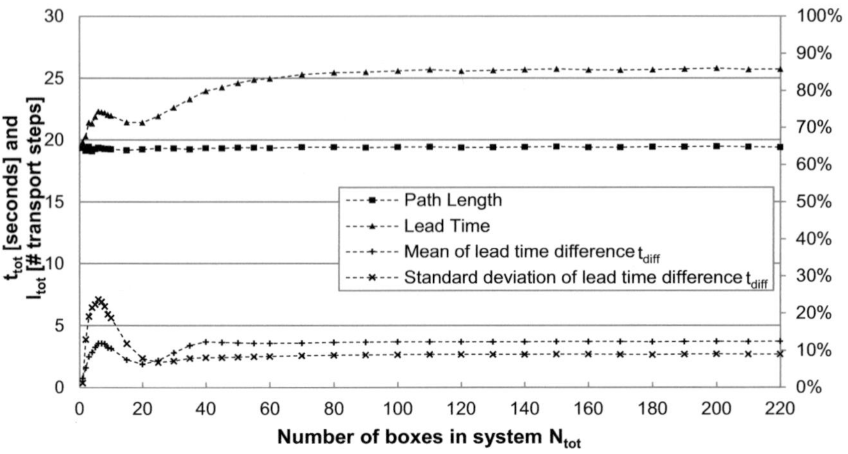

Figure 6.16: Lead time and path length and standard deviation of lead time difference under different system load with synchronization among sources

Figure 6.16 shows path length, lead time and leadtime difference between expected and real lead time for the standard layout with source synchronization. Comparing it to Figure 6.12, we can see that the expected lead time differs less from real lead time under high system loads. Lead time under a system load of 220 boxes is reduced by 4.8% leading to an increase of system throughput by 2.2%. The method of synchronized source modules only takes effect for system loads higher than 6 boxes. Therefore, lead time even decreases with increasing system load between 6 and 25 boxes. With system loads under 6 boxes, the interarrival time between boxes on any source is quite high. During these times, the logical clocks of source modules are not set forward which then leads to waiting times if the routes interfere with routes of boxes that have been routed much earlier.

The positive effect of synchronizing sources can also be observed in other layouts (group B and C of layout population of Table B.1). Figure 6.17 shows that the relative decrease of waiting time is proportional to the relative decrease of difference between expected lead time and real lead time. Figure 6.18 shows that the relative decrease of lead time tends to be higher if grid modules are less synchronized in systems without synchronization of sources.

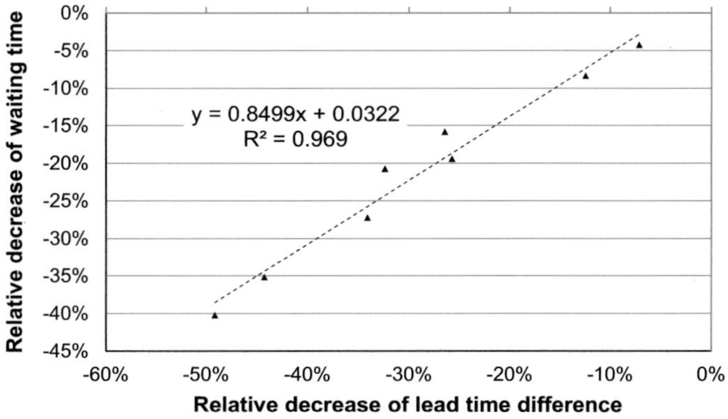

Figure 6.17: Relative decrease of waiting time as function of relative decrease of lead time difference

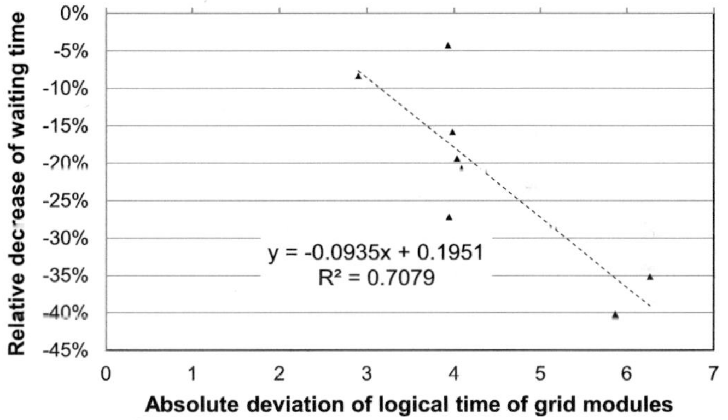

Figure 6.18: Relative decrease of waiting time as function of deviation of logical time of grid modules without source synchronization

In these 8 layouts, the lead time has been reduced between 2 and 13%. The number of boxes on the grid do not change which leads to an increase of throughput between 2 and 17% depending on the layout.

6.4 System Behavior in Different Layouts

In this section, we investigate the influence of layout characteristics on system behavior. In real world scenarios, it is desirable to maximize throughput while minimizing invest which is proportional to the number of modules in the system. Often, boundary conditions are given such as required system throughput or positioning of sources and destinations. We first investigate the impact of system size and number of sources in layouts with the same positioning of sources and destinations. Thereupon, we investigate the impact of positioning of sources and destinations on system throughput.

6.4.1 Impact of System Size and Number of Sources on Throughput

Figure 6.19 shows the throughput of 63 layouts out of the layout population all of which have source modules at the NORTH border and destination modules at the remaining three borders (positioning N-SEW). The layouts are all rectangular; some with "cropped" corners. The aspect ratio, the number of modules, sources and destinations vary. The layouts are divided in four groups according to the ratio of number of sources to number of modules (source-module-ratio). We can see that the throughput increases with increasing number of modules and increasing source-module-ratio because more modules and more sources facilitate more boxes on the grid. The results can be approximated with power functions (see dotted functions in Figure 6.19) with exponents around 0.45. We use the power function as approximation because of Little's law: If the number of modules is multiplied by factor x with stable aspect ratio, the lead time changes approximately with factor $\sqrt{x}$ because it is related to length and width of the layout. Therefore, throughput is nearly proportional to the square root of number of modules.

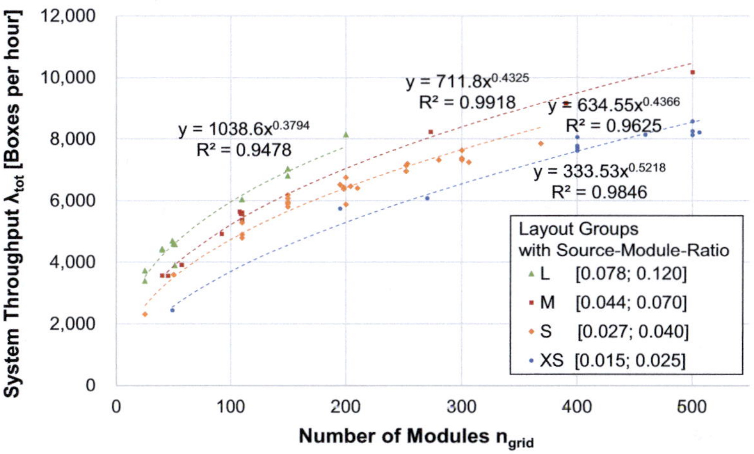

Figure 6.19: System performance of different layouts as a function of the number of modules and for different ratio of number of sources to number of modules

6.4.2 Impact of Positioning of Sources and Destinations on Throughput

Figure 6.20 shows the throughput of five layout groups under full load. The layouts of one layout group only differ in positioning of sources and destination modules: The first letters indicate the position of sources, the second letters the position of destinations. The graphical representation of the layouts can be found in the appendix in Figure B.1 and B.2.

Figure 6.20 shows the achieved throughput of all layouts and Figure 6.21 shows the results of analysis of variance for the different positionings of sources and destinations. For this analysis, we express the measures as portion of the maximal value in this layout group. Positioning EW-NS significantly outperforms the other positionings in the studied layout groups (highest mean with lowest deviation). Different hypotheses could help to explain throughput differences:

- In layouts with lots of opposing box traffic, throughput is lower, whereas throughput is higher in layouts with one or two main flow directions. This effect favors N-SEW and EW-NS positioning as indicated by a high number of boxes on the grid on the right side of Figure 6.21.

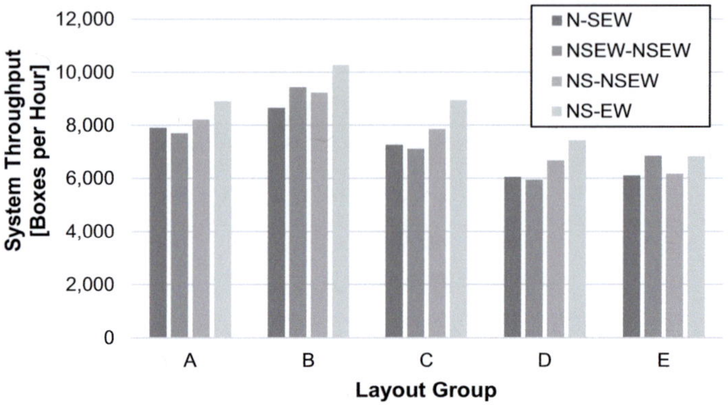

Figure 6.20: Throughput of different layout groups with differing positioning of sources and destinations under full load

Positioning of sources and destinations	Throughput μ	Throughput σ²	Shortest path length between source and destination μ	Shortest path length between source and destination σ²	Number of boxes on grid μ	Number of boxes on grid σ²
N-SEW	84.96%	0.1458%	99.11%	0.0395%	85.13%	1.7704%
NSEW-NSEW	87.60%	0.7406%	84.94%	0.0353%	73.61%	1.2075%
NS-NSEW	89.98%	0.0244%	83.71%	0.0279%	72.73%	0.6531%
EW-NS	99.94%	0.0002%	95.48%	0.1180%	99.43%	0.0165%

	Throughput F	Throughput p-value	Shortest path length F	Shortest path length p-value	Number of boxes on grid F	Number of boxes on grid p-value
	9.40	0.0008	53.05	0.0000	8.55	0.0013

Figure 6.21: Analysis of variance for differing positioning of sources and destinations under full load

- In layouts with shorter path lengths, lead time is lower and therefore throughput is higher. This effect favors NSEW-NSEW and NS-NSEW layouts as indicated by the short distance between source and destination in the middle of Figure 6.21.
- Uniform occupancy and relative throughput suggest that no bottleneck exists and throughput is higher. In section 6.2.1, we have seen that the rows close to source modules represent a bottleneck in N-SEW layouts.

Looking at the results in Figure 6.21, opposing traffic reduces the occupancy of the grid strongly. However, NSEW-NSEW and NS-NSEW layouts perform quite well thanks to the short average path length. EW-NS layouts achieve highest average throughput despite relatively long path lengths because they achieve a high occupancy and do not have a bottleneck like N-SEW layouts.

For application of GridSorter in real world scenarios, it is likely that positioning of sources and destinations is given by the customer. In addition, it is relevant how much traffic from one specific source to one specific destination is required. In our simulation studies, every destination is equally assigned to boxes from all sources. Therefore, the positioning of sources and destinations is not further investigated.

6.4.3 Communication Effort

In this section, we study the communication effort in our layout population to draw conclusion whether the system could realize the described routing algorithm on real hardware. Figure 6.22 shows the average number of messages for one reservation process. Since the transport process only starts when the reservation process is finished, a high number of reservation messages could hinder the box from being transported. Since the box occupies the source module while waiting, system throughput is reduced.

In Figure 6.22, we see that the number of messages per reservation increases with the average shortest path length between source and destination: The longer the path, the more messages need to be sent. The trend function shows that the number of reservation messages increases potentially with the average shortest path length. The longer the distance between source and destination, the more messages need to be sent. It is not a linear trend function because with increasing path length, it is more likely that alternative routes need to be explored. If sources are direct neighbors such as in layout 65, 67 and 79, more reservation messages are needed. One reason could be that running path searches interfere more with each other because source nodes are positioned close to each other. We conclude that the size of the layout influences the communication effort indirectly. The distance between sources and destinations is the determining measure and it is not only dependent on the size but also in the shape of the layout.

The highest communication effort is required in layout 79 and 51 where around 340 messages are sent in average per reservation. If it takes around 3 ms to send a message from one module to another, a reservation process of 340 messages takes around 1 second in average which is acceptable since it corresponds to the duration of one transport step. Nevertheless, all reservation processes with more reservation messages affect system throughput. Therefore, we study the effect of partial route reservation in the next section.

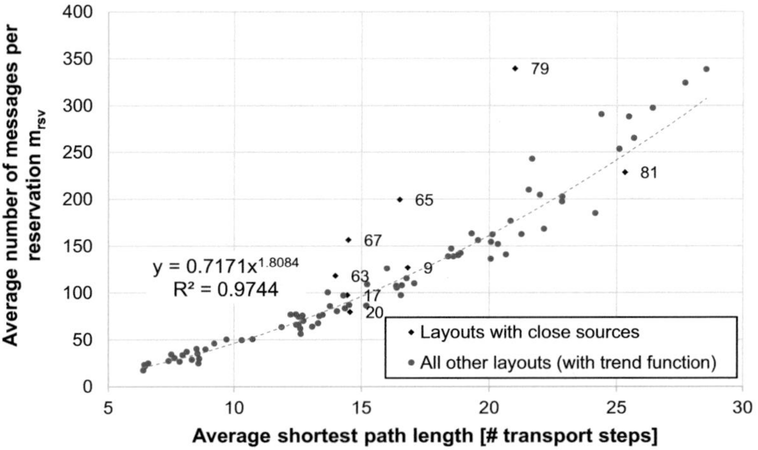

Figure 6.22: Communication effort under full load dependent on average shortest path length (layout IDs next to some points)

6.5 Partial Route Reservation

In this section, we study whether partial route reservation increases system throughput by reducing the number of messages before the transport of a box starts. Again, we use our population of layouts and set the number of boxes in the system so high that there are always boxes waiting for sources to get available. In contrast to previous sections, the duration of message transfer is now set to 3 milliseconds to take into account the effect of long reservation processes.

The parameters for partial route reservation have been set as following (cf. section 5.3.4): The remaining distance from the inter-destination module should be less than half of the distance between source and destination. Consequently, more than half of the path should be reserved. A source module starts the reservation if the number of sent messages exceeds the message count limit of 100 messages. These settings have shown best results (cf. Appendix C).

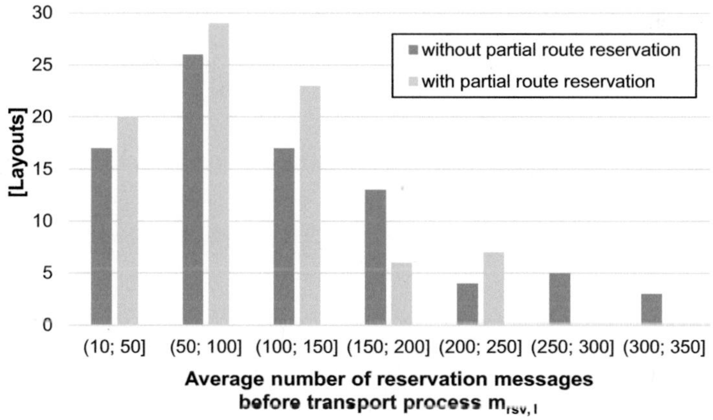

Figure 6.23: Average number of reservation messages in different layouts with and without partial route reservation

Figure 6.23 illustrates the average number of reservation messages before the transport process starts in systems with and without partial route reservation. Without partial route reservation, this measure corresponds to the average number of messages per reservation process. The bars indicate in how many layouts the average number of messages lies within the interval on the x-axis. We see clearly that the average number of messages before start of transport is reduced by partial route reservation. With partial route reservation, it is less than 250 messages in average in all layouts. Even though this is positive for system functionality, we now have to investigate if system throughput is influenced negatively.

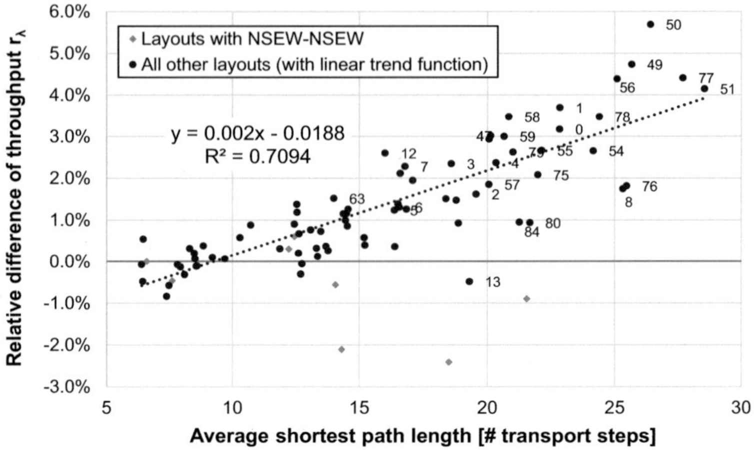

Figure 6.24: Relative difference of throughput in different layouts due to partial route reservation (layout IDs next to some points)

Figure 6.24 shows the relative difference of system throughput r_λ for all 85 layouts. It is defined by

$$r_\lambda = \frac{\lambda_{with} - \lambda_{without}}{\lambda_{without}}$$

where λ_{with} is throughput with partial route reservation and $\lambda_{without}$ is throughput without partial route reservation. The x-axis indicates the average shortest path length between source and destination. It becomes clear that partial route reservation has a positive effect on throughput in most layouts. For lots of layouts with small distance between source and destination, the difference is smaller than 1% which is in magnitude of the standard error and thus not significant. The positive effect on throughput increases with the average shortest path length between source and destination and can be approximated with a linear trend function. The reason is that communication effort increases with increasing path length. The negative effect of reservation processes with lots of messages is reduced by partial route reservation. Two layouts exist where partial route reservation has a clearly negative impact. Since both layouts show NSEW-NSEW positioning, the reduced quality of the partially reserved route seems to be important with opposing box traffic.

6.6 System Behavior under Varying Transport Times

All simulation experiments in previous sections have been conducted under fixed transport times meaning that each transport from one module to the next one takes exactly 1 second. In reality, acceleration and deceleration influence transport times. Consequently, the transport from one module to the next one takes shorter if the box does not need to stop but can move through several modules. In addition, right-angle-transfer module have different transport times depending on the conveying direction because belts have more friction. We have measured transport times on a 5×5-demonstrator system of GridSorter. Transport time in direction of belts varies between 0.6 and 0.85 seconds. In direction of rollers, it varies between 0.9 and 1.95 seconds. Since the variation of transport time origins from acceleration and deceleration, we modeled the physical behavior of transport of boxes depending on the number of movements without stop. In addition, the transport times of two boxes are adjusted if they are transported in tandem.

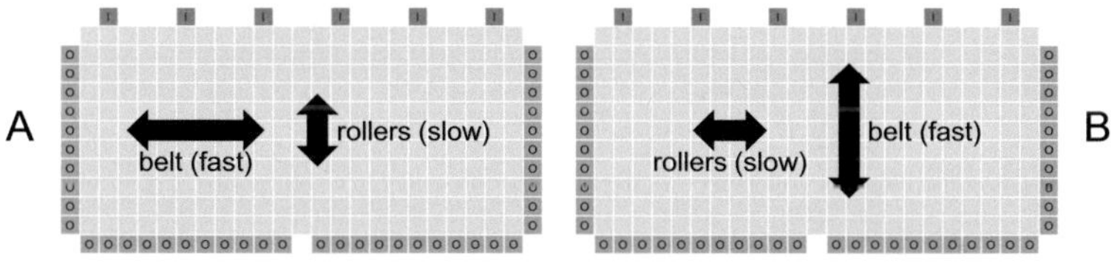

Figure 6.25: Two configurations of arrangement of conveyor modules in standard layout

We now study throughput and average transport times under varying transport times for the standard layout of Figure 6.2. The modules are arranged in configurations A and B (see Figure 6.25). In Figure 6.26, we can see that the same standard layout achieves more throughput in configuration A than in configuration B. We conclude that the faster conveying direction should be used for the longer side of the layout. Figure 6.26 indicates that a transport in direction of rollers takes around 0.97 seconds in averages, and in direction of belts around 0.7 seconds in average. Under low system load, transport times are a little shorter because boxes have to stop less often. We also see that average transport times do not differ much in both configurations.

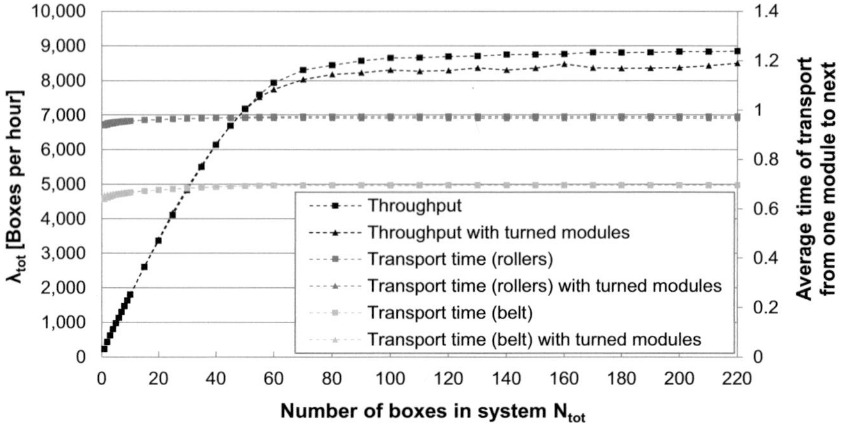

Figure 6.26: Throughput and average transport times under varying transport times for standard layout

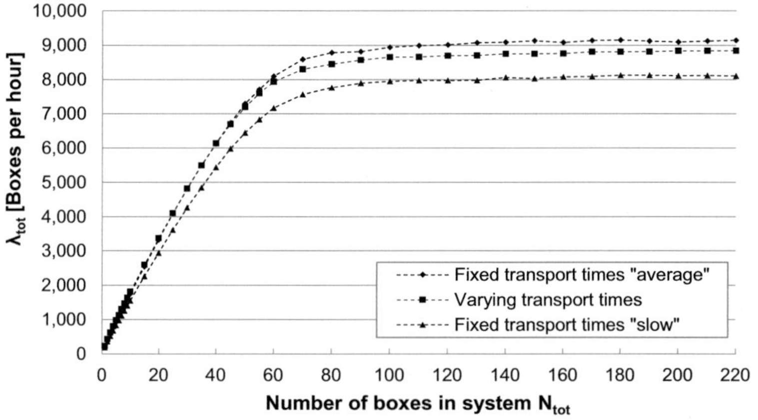

Figure 6.27: Throughput under varying and fixed transport times for standard layout

In order to assess the influence of varying transport times on throughput, we compare a system with varying transport times to a system with fixed transport times. Fixed transport times have been simulated with two different settings: With the setting "slow" we assume that a box stops after each transport step. With the setting "average", we assume that each transport step takes the average transport time from the experiment with varying transport times. Please note that transport times still differ in the conveying directions even with fixed transport times.

Figure 6.27 shows throughput under varying and fixed transport times for the standard layout with configuration A of Figure 6.25. Under low system loads, the system with varying transport times achieves the same throughput like the system with fixed transport times "average". Under high system load, the system with fixed transport times performs better. Because of varying transport times, boxes have to wait more often for each other. Nevertheless, it makes no sense to force the system to stop boxes after each transport. The system with fixed transport times "slow" achieves less throughput under all system loads.

The simulation experiments of this section show that the system control with logical time is able to handle varying transport times. Like this, the positive effect on system performance by transporting boxes through several movements without a stop can be exploited.

6.7 Conclusion on System Behavior Controlled with Logical Time

System behavior has been studied with GridSorter as showcase system. Most important findings are:

- With the described system control, the system does not show a decrease of throughput because of overload. It is thus not necessary to regulate system load.
- Uniform module occupancy and uniform relative module throughput have a positive effect on system throughput.
- The basic control parameters should be chosen in such way that routing possibilities are limited but not strictly.
- Synchronicity of logical clocks increases with increasing system load. Especially under low system loads, asynchronous logical clocks lead

to waiting times. With an additional synchronization among source modules, lead time can be decreased and system throughput increased.

- Related to layout characteristics, system size is most important for throughput. The ratio of sources to modules should be high enough. Layouts with EW-NS positioning outperform layouts with other positioning of same size and shape.
- Communication effort increases potentially with distance between source and path length. Therefore, system size influences communication effort indirectly.
- Partial route reservation shows positive impact both on the number of required messages before start of transport as well as on throughput. The positive effect on throughput increases linearly with distance between source and destination.
- Because of logical time, our system is robust to varying transport times that arise for example from acceleration and deceleration. Transporting boxes through several modules without a stop improves system performance.

The simulation studies in this section have mainly been focused on theoretical questions. For application of GridSorter, studies with real-world-requirements should be undertaken. Since system throughput depends on many different control parameters and layout characteristics, it is necessary to conduct simulation experiments for each application case. Therefore, further investigation should have the objective to generate layouts automatically (Fechtig 2014).[2]

[2] A similar idea has been presented for Cognitive Conveyors by Shchekutin et al. (2015).

7 Conclusion

Everyone is ignorant,
only on different subjects.
-- Will Rogers

The fourth industrial revolution aims to transform production systems by using numerous, small-scaled electronic devices. By sharing information and taking independent decisions, the systems promise more flexibility and robustness. In order to profit from these opportunities, new decentralized control algorithms have to be developed. This thesis makes a contribution to decentralized control of modular material handling systems: Logical time is, for the first time, used as control principle for routing of boxes and deadlock-free transport to their destinations. The results look promising.

7.1 Conclusion on Thesis

What has been achieved in this thesis? Have all research questions been answered?

In Chapter 2, we first have given an overview of existing modular material handling systems with decentralized control. These systems have been classified according to different criteria focusing routing and deadlock handling. From this classification, the research gap has been identified: No control algorithm exists for the transportation of boxes that prevents deadlocks and is generally applicable. In addition, time-window-based route reservation is not used in any of these systems. The GridSorter has been chosen as showcase system because it shows high risk of deadlocks.

Take-Home-Message of Chapter 2 Multiple different modular material handling systems have been developed and result in some very impressive control algorithms. Nevertheless, no deadlock prevention algorithm exists that can be applied universally.

In Chapter 3, the principle of logical time has been successfully transferred to the control of modular material handling systems. Time-window-based routing has been chosen as routing strategy and the principle of logical time has been adapted to our application. It has been proven that the system is deadlock-free. In addition, the system is robust against varying transport times.

> **Take-Home-Message of Chapter 3** Logical time is a very structured control principle that helps in designing deadlock-free control algorithms.

In Chapter 4, the control architecture has been introduced. The two components *reservation manager* and *transport manager* have been explained in detail because they are responsible for route reservation and release of transport steps. As path-finding algorithm, IDA* has been adapted to decentralized control and time-dependent route reservation. We call the developed algorithm DIDA* standing for *Decentralized Iterative Deepening A**. Using time-dependent route reservation, the size of the search tree increases. Scalability of the control algorithm can be achieved by selecting inter-destination modules and thus accepting partial route reservation. The transport manager is responsible for guaranteeing that the execution of transport steps corresponds to the reserved timestamps.

> **Take-Home-Message of Chapter 4** Iterative Deepening A* can be used as path-finding algorithm for time-dependent route reservation in modular material handling systems.

In Chapter 5, the simulation model of GridSorter has been presented. Compared to reality, simplifications are required in order to achieve acceptable simulation effort. All input parameters that can be changed and different important output variables that can be used as performance indicators have been presented.

> **Take-Home-Message of Chapter 5** In order to study system behavior of material handling systems with decentralized control, an agent-based simulation model should be implemented.

The objective of Chapter 6 was to make system behavior better understandable. Therefore, results of different simulation studies have been presented. For improving route reservation, it is helpful to define limitations such as the limitation of path length that reduce routing possibilities, but still offer enough of them. Two characteristics of layouts influence system throughput mainly: the number of modules i.e. system size and the number of sources related to system size. Synchronicity of logical clocks increases with increasing system load. For low system load, synchronicity depends strongly of positioning of sources and sinks.

Take-Home-Message of Chapter 6 Asynchronous logical clocks can lead to waiting times. With an additional synchronization among source modules, lead time can be decreased and system throughput increased.

In this thesis, a control principle with logical time for material handling systems with decentralized control has been developed, theoretically proven and tested in simulation. In addition, it has been successfully implemented on a 5×5-demonstration system of GridSorter.

7.2 Outlook

What else can be achieved with logical time? Which further system functionalities are imaginable? How could the principle of logical time be applied to other systems?

Related to the presented control algorithm with logical clocks, a multitude of questions still remains after closing this thesis:

- How could a better synchronicity of all logical clocks be achieved by not adding too much communication effort? How much does a better synchronicity increase system throughput?
- One advantage of logical time is that the system is robust against different and varying transport times respectively. Nevertheless, waiting times could occur due to reserving time-windows in logical time that differ a lot to physical time. Could waiting times be reduced by a smaller scale of logical timestamps? How would this impact communication effort for route reservation?
- What kind of local information such as module occupancy can be additionally included in routing decisions to improve the quality of the found path?

Not only the presented algorithm could be improved, but further system functionalities could be achieved by enhancing the presented route reservation process. The partial ordering of transport steps incorporated in the principle of logical time could for example be used for buffering (cf. Sohrt et al. (2014)) or sequencing of boxes. In the case of sequencing, the boxes must leave the system through a defined destination in a defined sequence.

Furthermore, we believe that logical time could facilitate routing and deadlock handling in many other systems such as those presented in Chapter 2. The transfer of logical time to systems with high density of goods such as GridStore and to shuttle systems such as Store Biter is a challenging task. In addition, logical time could also be used in production for routing of items through multiple production steps.

Finally, nothing remains but questions looking into the future: Which of the systems presented in Chapter 2 will be industrialized next? When is the first GridSorter going to be used in production or logistics? We are excited to experience further progress of the announced fourth industrial revolution. In our opinion, it is a huge challenge to bring together people who are working on different research and development projects so that one big revolution emerges from the multitude of small innovations.

Notation

The notation glossary is divided into multiple parts. The right column indicates the page number of the introduction of the notation.

Legend for Graphical Representation of Figures with Modules

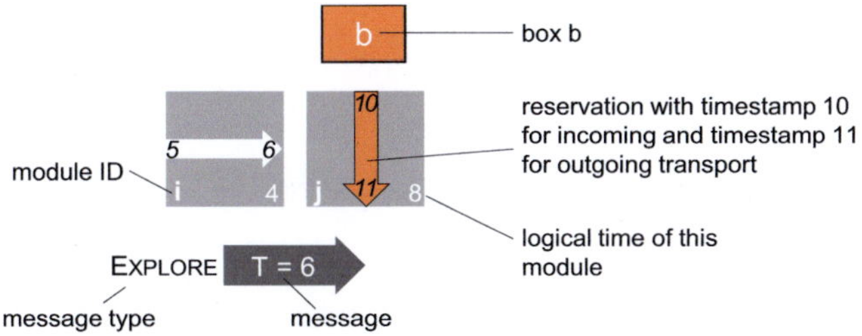

Figure 7.1: Legend for graphical representation of figures with modules

Resources and Processes

R_i	Resource i	39
R	Set of all resources in the system	56
P_a	Process a	39
P	Set of all processes in the system	56

Logical Times and Clocks and Timestamps

C_i	logical clock of process or resource i	48
C	set of all logical clocks in the system	48
$C\langle a \rangle$	logical time assigned to event a by the entire set of logical clocks	48

$C_i\langle a\rangle$	logical time assigned to event a by logical clock of process or resource i	48
T	logical timestamp	49
T_{in}	timestamp for incoming transport defining the start of reservation	51
T_{out}	timestamp for outgoing transport defining the start of reservation	51
T_{msg}	timestamp included in reservation message	77
T_{prop}	alternative timestamp proposed for transport	77

FlexPorts

f_{in}	incoming port	79
f_{out}	outgoing port	79
F_{all}	set of all connected flexports of one module	84
F_{elg}	set of connected flexports eligible as outgoing port for a specific reservation	84

Path Costs: Path Length and Lead Time

t_{max}	maximal lead time	74
t_{exp}	expected lead time	73
t_{up}	upstream lead time	74
t_{real}	real lead time from source to destination	73
t_{diff}	difference between expected and real lead time	110
$l_{dest}(f_i)$	minimal path length to destination through port f_i measured in transport steps	74
l_{tot}	path length from source to destination measured in transport steps	110

Input Parameters and Performance Measures

n_{grid}	number of grid modules in a layout	105
λ_{tot}	system throughput: number of boxes reaching their destination per time interval	109

λ_i	throughput of module i	110
N_{grid}	number of boxes on the grid including the source and destination modules	109
N_{wait}	number of boxes waiting for source modules to get available	109
N_{tot}	number of boxes in the system (sum of N_{grid} and N_{wait})	109
o_i	occupancy of module i	110
$\bar{C}(t)$	average logical time at physical time t	109
$d_{abs,g}$	Average absolute deviation of logical time of one module from mean logical time of group g of modules	109
$d_{g,h}$	Average deviation of mean logical time of group g from mean logical time of group h	109
m_{rsv}	number of messages per reservation	110
$m_{rsv,1}$	number of messages per reservation that are sent before the first transport starts	110

Acronyms

AGV	Autonomous Guided Vehicle
CONWIP	Constant work in process
GC	Granting Condition
IDA*	Iterative-Deepening A*
MHS	Material Handling System
msg	message
rsv	reservation

References

Alt, K. (2014). *Erweiterung und Optimierung des Modells und der Steuerung eines dezentral gesteuerten Stetigfördersystems.* Master Thesis supervised by, Z. Seibold, Karlsruhe Institute of Technology. Karlsruhe.

Broy, M., M. Cengarle and E. Geisberger (2012). Cyber-Physical Systems: Imminent Challenges. In: R. Calinescu and D. Garlan (eds.), *Large-Scale Complex IT Systems. Development, Operation and Management*, Volume 7539 of *Lecture notes in computer science*, p. 1–28. Springer Berlin Heidelberg.

Chandy, K. M., J. Misra and L. M. Haas (1983). Distributed deadlock detection. *ACM Transactions on Computer Systems 1*(2), p. 144–156.

Chisu, R., F. Kuzmany and W. A. Günthner (2010). Realisierung einer agentenbasierten Steuerung für Elektrohängebahnsysteme. In: W. A. Günthner and M. ten Hompel (eds.), *Internet der Dinge in der Intralogistik*, VDI-Buch, p. 263–274. Springer Berlin Heidelberg.

Coffman, E. G., M. Elphick and A. Shoshani (1971). System Deadlocks. *ACM Comput. Surv. 3*(2), p. 67–78.

Cormen, T. H. (2009). *Introduction to algorithms* (3rd ed. ed.). Cambridge, Mass.: MIT Press.

Dijkstra, E. W. (1959). A note on two problems in connexion with graphs. *Numerische Mathematik 1*(1), p. 269–271.

Elger, J., C. Haußner, M. Hofmeister and G. Baier (2010). Chancen und Herausforderungen von dezentral gesteuerten Flughafen-Gepäckförderanlagen. In: M. ten Hompel (eds.), *Internet der Dinge in der Intralogistik*, VDI, p. 275–294. Heidelberg [u.a.]: Springer.

Emmerich, J. S., M. Roidl, T. Bich and M. ten Hompel (2012). Entwicklung von energieautarken, intelligenten Ladehilfsmitteln am Beispiel des inBin. *Logistics Journal Proceedings.*

Fechtig, A. (2014). *Automatisierte Layoutgenerierung für ein modulares Sortiersystem mithilfe naturinspirierter Algorithmen.* Master Thesis supervised by, Z. Seibold, Karlsruhe Institute of Technology. Karlsruhe.

Feldhorst, S., M. ten Hompel and M. Fiedler (2010). Paket Royale - Dezentrale Steuerung für das Internet der Dinge. *Logistics Journal Proceedings.*

Festo AG & Co. KG (06.10.2015). Mit Motion Cube zur individualisierten Massenfertigung: Außergewöhnliches Entwicklungsprojekt gewinnt Handling-Award 2015.

flexlog GmbH (2015). Dezentral steuerbar – Modulbaukasten mit Flex-Technology, `www.flexlog.de/de/produkte/flextechnology`. [Online] (Accessed on 23.07.2015).

Freund, P. (2015). *Auswertung GridSorter.* Seminar Paper supervised by, Z. Seibold, Karlsruhe Institute of Technology. Karlsruhe.

Furmans, K., F. Schönung and K. R. Gue (2010). Plug-and-Work Material Handling Systems. In: MHIA (eds.), *Proceedings of the International Material Handling Research Colloquium (IMHRC).*

Fuß, B., Z. Seibold and K. Furmans (2015). Leistungsverfügbarkeit eines modularen Sorters. In: M. ten Hompel (eds.), *Tagungsband des Symposium Leistungsverfügbarkeit in der Logistik (im Druck).* Logistics Journal.

Gawrilow, E., E. Köhler, R. H. Möhring and B. Stenzel (2008). Dynamic Routing of Automated Guided Vehicles in Real-time. In: H.-J. Krebs and W. Jäger (eds.), *Mathematics – Key Technology for the Future*, p. 165–177. Springer Berlin Heidelberg.

Gebhardt Fördertechnik GmbH (2015). StoreBiter 500-OLPS - Vorteile, `http://www.gebhardt-foerdertechnik.de/de/produkte/lagertechnik/shuttlesysteme/paletten-storebiter-500-olps/storebiterr-500-olps-vorteile.html`. [Online] (Accessed on 21.04.2015).

Geier, B. (2015). *Klassifizierung von dezentral gesteuerten Materialflusssystemen.* Seminar Paper supervised by, Z. Seibold, Karlsruhe Institute of Technology. Karlsruhe.

Gue, K. R., K. Furmans, Z. Seibold and O. Uludag (2014). GridStore: A Puzzle-Based Storage System With Decentralized Control. *IEEE Transactions on Automation Science and Engineering 11*(2), p. 429–438.

Gue, K. R., O. Uludag and K. Furmans (2012). A High-Density System for Carton Sequencing. In: Bundesvereinigung für Logistik (eds.), *Proceedings of the 6th International Scientific Symposium on Logistics (ISSL).*

Hama, K., S. Mikami, K. Suzuki and Y. Kakazu (2002). Motion coordination algorithm for distributed agents in the cellular warehouse problem.

Artificial Life and Robotics 6(1-2), p. 3–10.

Hart, P., N. Nilsson and B. Raphael (1968). A Formal Basis for the Heuristic Determination of Minimum Cost Paths. *IEEE Transactions on Systems Science and Cybernetics 4*(2), p. 100–107.

Haude, J. (2014). *Synchronisierung der Fördermodule eines dezentral gesteuerten Sorters.* Bachelor Thesis supervised by, Z. Seibold, Karlsruhe Institute of Technology. Karlsruhe.

Hofmeister, M., G. Baier and M. Gärtner (2010). Strategien für die dezentrale agentenbasierte Steuerung von Materialflusssystemen. In: M. ten Hompel (eds.), *Internet der Dinge in der Intralogistik*, VDI, p. 119–140. Heidelberg [u.a.]: Springer.

Kagermann, H., W. Wahlster and J. Helbig (2012). Umsetzungsempfehlungen für das Zukunftsprojekt Industrie 4.0–Abschlussbericht des Arbeitskreises Industrie 4.0. *Forschungsunion im Stifterverband für die Deutsche Wissenschaft. Berlin.*

Kim, C. W. and J. M. A. Tanchoco (1991). Conflict-free shortest-time bidirectional AGV routeing. *International Journal of Production Research 29*(12), p. 2377–2391.

Korf, R. E. (1985a). Depth-first iterative-deepening: An optimal admissible tree search. *Artificial Intelligence 27*(1), p. 97–109.

Korf, R. E. (1985b). Iterative-deepening-A: An Optimal Admissible Tree Search. In: A. Joshi (eds.), *Proceedings of the Ninth International Joint Conference on Artificial Intelligence*, Los Altos, Calif., p. 1034–1036. Kaufmann.

Korf, R. E. (1990). Real-time heuristic search. *Artificial Intelligence 42*(2-3), p. 189–211.

Krühn, T. (2014). *Dezentrale, verteilte Steuerung flächiger Fördersysteme für den innerbetrieblichen Materialfluss.* Dissertation, Gottfried Wilhelm Leibniz Universität Hannover. Hannover.

Krühn, T., M. Radosavac, N. Shchekutin and L. Overmeyer (2013). Decentralized and Dynamic Routing for a Cognitive Conveyor. In: IEEE/ASME (eds.), *Proceedings of the International Conference on Advanced Intelligent Mechatronics (AIM)*, p. 436–441.

Lamport, L. (1978). Time, Clocks, and the Ordering of Events in a Distributed System. *Communications of the ACM 21*(7), p. 558–565.

Lanfer Automation GmbH & Co. KG (2015). THINGtelligence, `www.`

`thingtelligence.de`. [Online] (Accessed on 23.07.2015).

Levine, G. N. (2005). The classification of deadlock prevention and avoidance is erroneous. *ACM SIGOPS Operating Systems Review 39*(2), p. 47–50.

Little, J. D. C. (1961). A Proof for the Queuing Formula: $L = \lambda W$. *Operations Research 9*(3), p. 383–387.

Lynch, N. A. (1981). Upper bounds for static resource allocation in a distributed system. *Journal of Computer and System Sciences 23*(2), p. 254–278.

Mayer, S. (2009). *Development of a Completely Decentralized Control System for Modular Continuous Conveyor Systems.* Dissertation, Universität Karlsruhe.

Mayer, S. and K. Furmans (2010). Deadlock Prevention in a Completely Decentralized Controlled Materials Flow Systems. *Logistics Research 2*(3-4), p. 147–158.

Maza, S. and P. Castagna (2005a). A performance-based structural policy for conflict-free routing of bi-directional automated guided vehicles. *Computers in Industry 56*(7), p. 719–733.

Maza, S. and P. Castagna (2005b). Sequence based hierarchical conflict-free routing strategy of bi-directional automated guided vehicl. In: P. Zítek (eds.), *World Congress*, IFAC proceedings volumes, p. 2050. IFAC, Elsevier.

Montreuil, B., R. D. Meller and C. Thivierge (2012). Functional Design of Physical Internet Facilities: A Road-Based Crossdocking hub. In: MHIA (eds.), *Proceedings of the International Material Handling Research Colloquium (IMHRC).*

Oberkersch, W. (2013). *Entwicklung eines zeitfensterbasierten Algorithmus für ein dezentral gesteuertes Stetigfördersystem.* Masterarbeit supervised by, Z. Seibold, Karlsruhe Institute of Technology. Karlsruhe.

Qiu, L., W.-J. Hsu, S.-Y. Huang and H. Wang (2002). Scheduling and routing algorithms for AGVs: A survey. *International Journal of Production Research 40*(3), p. 745–760.

Raynal, M. (2013). *Distributed algorithms for message-passing systems.* Berlin and New York: Springer.

Roidl, M. (2012). *Agentifizierung der Intralogistik.* Forschungsbericht, Technische Universittät Dortmund.

Schwab, M. (2015). *A decentralized control strategy for high density material flow systems with automated guided vehicles.* Dissertation, Karlsruhe Institute of Technology. Karlsruhe.

Seibold, Z. and K. Furmans (2014). GridSorter - Logische Zeit in dezentral gesteuerten Materialflusssystemen. *Logistics Journal.*

Seibold, Z. and K. Furmans (2016). Plug&Play-Fördertechnik in der Industrie 4.0. In: B. Vogel-Heuser, T. Bauernhansl, and M. ten Hompel (eds.), *Handbuch Industrie 4.0*, p. 1–17. Berlin, Heidelberg: Springer Berlin Heidelberg.

Seibold, Z., K. Furmans and M. Gebhardt (2015). Steuerung dezentraler Materialflusssysteme mittels Logischer Zeit. *Vortrag auf der VDI-Fachkonferenz Shuttle in der Logistik.*

Seibold, Z., M. Gebhardt and T. Stoll (2014). Mehr Nutzen mit dem GridSorter. *Hebezeuge Fördermittel* (2014 - 5), p. 260–262.

Seibold, Z., T. Stoll and K. Furmans (2013). Layout-optimized sorting of goods with decentralized controlled conveying modules. In: IEEE (eds.), *Proceedings of the 7th Annual Systems Conference (SysCon)*, p. 628–633.

Shchekutin, N., L. Overmeyer, S. Zobnin and V. Shkodyrev (2015). Mathematical methods for the configuration of transportation systems with focuson continuous and modular matrix conveyors. *Logistics Journal.*

Silver, D. (2005). Cooperative Pathfinding. In: AAAI (eds.), *Proceedings of Conference on Artificial Intelligence and Interactive Digital Entertainment (AIIDE)*, p. 117–122.

Sittivijan, P. (2014). *Modular Warehouse Control: Simultaneous Rectilinear Movement of Multiple Objects within a Limited Free Space Environment.* North Carolina: North Carolina State University.

Sohrt, S., Z. Seibold, T. Krühn, L. Prössdorf, L. Overmeyer and K. Furmans (2014). Buffering Algorithms for Modular, Decentralized Controlled Material Handling Systems: 1st Symposium on Automated Systems and Technologies (AST). *Berichte aus dem ITA, Garbsen: TEWISS-Technik und Wissen GmbH 2014* (4), p. 29–36.

Stenzel, B. (2008). *Online Disjoint Vehicle Routing with Application to AGV Routing.* Dissertation, Technische Universität Berlin. Berlin.

Tadakuma, K., R. Tadakuma, K. Ioka, T. Kudo, M. Takagi, Y. Tsumaki, M. Higashimori and M. Kaneko (2012a). Additional manipulating function for limited narrow space with omnidirectional driving gear. In:

IEEE/RSJ (eds.), *2012 IEEE/RSJ International Conference on Intelligent Robots and Systems (IROS 2012)*, p. 5438–5439.

Tadakuma, K., R. Tadakuma, K. Ioka, T. Kudo, M. Takagi, Y. Tsumaki, M. Higashimori and M. Kaneko (2012b). Omnidirectional driving gears and their input mechanism with passive rollers. In: IEEE/RSJ (eds.), *2012 IEEE/RSJ International Conference on Intelligent Robots and Systems (IROS 2012)*, p. 2881–2888.

Tanenbaum, A. S. (2011). *Computer networks* (5th ed. ed.). Boston: Pearson Education.

Tanenbaum, A. S. and H. Bos (2015). *Modern operating systems* (Fourth edition, Global edition ed.). Boston, [Massachusetts]: Pearson.

ten Hompel, M. (eds.) (2010). *Internet der Dinge in der Intralogistik.* VDI. Heidelberg [u.a.]: Springer.

Tenerowicz-Wirth, P. (2013). *Kommunikationskonzept für selbststeuernde Fahrzeugkollektive in der Intralogistik.* München: Universitätsbibliothek der TU München.

ter Mors, A. (2009). *The world according to MARP: Multi-agent route planning*, Volume 2010-11 of *SIKS Dissertation Series.* [S.l.]: [s.n.].

ter Mors, A., X. Mao, J. Zutt, C. Witteveen and N. Roos (2008). Robust Reservation-Based Multi-Agent Routing. In: M. Ghallab (eds.), *ECAI 2008*, Volume vol. 178 of *Frontiers in artificial intelligence and applications*, Amsterdam ... [etc.], p. 929–930. IOS Press.

Uludag, O. (2014). *GridPick: A High Density Puzzle Based Order Picking System with Decentralized Control.* Dissertation, Auburn University. Auburn, Alabama.

Uriarte, C., H. Thamer and M. Freitag (2015). Fördertechnik aus der Zelle: Hochflexibles Fördersystem aus kommunizierenden und kooperierenden Modulen. *Hebezeuge Fördermittel 2015*(10).

Vivaldini, K. C., L. Rocha, M. Becker and A. P. Moreira (2015). Comprehensive Review of the Dispatching, Scheduling and Routing of AGVs. In: A. P. Moreira, A. Matos, and G. Veiga (eds.), *CONTROLO'2014 – Proceedings of the 11th Portuguese Conference on Automatic Control*, Volume 321 of *Lecture Notes in Electrical Engineering*, p. 505–514. Springer International Publishing.

Wilke, M. (2006). *Wandelbare automatisierte Materialflusssysteme für dynamische Produktionsstrukturen.* Dissertation, Technische Universität München. München.

Yalcin, A., O. Schocke and A. Koberstein (2015). Gitterbasierte Lagersysteme – neue Lösungen für Frühgepäckspeicher: Bewegungen auf kleiner Fläche. *Hebezeuge Fördermittel 2015*(8), p. 386–389.

Zutt, J., A. van Gemund, M. de Weerdt and C. Witteveen (2010). Dealing with Uncertainty in Operational Transport Planning. In: R. R. Negenborn, Z. Lukszo, and H. Hellendoorn (eds.), *Intelligent Infrastructures*, Volume 42 of *Intelligent Systems, Control and Automation: Science and Engineering*, p. 349–375. Springer Netherlands.

List of Figures

List of Tables

A Decentralized Control

Content	**Rsv msg**	**Rsv entry**	**Tsp msg**	**Comment**
Reservation ID	x	x	x	
Destination ID	x	x		General information set by source
Source ID	x	x		
Maximal lead time t_{max}	x	x		
Reservation msg type	x			
Transport msg type	x		x	Node information updated by each module
Timestamp T_{msg}	x		x	
Expected lead time t_{exp}	x			
Port permission	x			Indicating the BACKTRACK cause
Proposed timestamp T_{prop}	x			
Reservation state		x		
Transport state		x		
Incoming timestamp T_{in}		x		
Outgoing timestamp T_{out}		x		Description of Reservation
Incoming port f_{in}		x		
Outgoing port f_{out}		x		
Upstream lead time t_{up}		x		
Expected lead time for each port $t_{exp}\,(F_{all})$		x		Routing Information
Outgoing timestamp for each port $T_{out}\,(F_{all})$		x		
Permission for each port		x		

Table A.1: Content of Reservation Message and Table Entry and Transport Message

B Layouts

In the following, two populations of layouts are introduced. Each population is described with an overview of the layout characteristics and is then followed by a graphical representation of each layout included in the population

Layout Group	Position of sources and destinations	# sources	# destinations	# modules	# rows	# columns	Shape	Aspect ratio	Source-destination-ratio	Source-module-ratio	Distance from sources to destinations
A	N-SEW	6	42	253	11	23	rectangular	2.09	7.0	0.024	19.206
	EW-NS	6	42	253	11	23	rectangular	2.09	7.0	0.024	18.000
	NSEW-NSEW	6	42	253	11	23	rectangular	2.09	7.0	0.024	15.841
	NS-NSEW	6	42	253	11	23	rectangular	2.09	7.0	0.024	15.746
B	N-SEW	10	42	253	11	23	rectangular	2.09	4.2	0.040	18.905
	EW-NS	10	42	253	11	23	rectangular	2.09	4.2	0.040	18.000
	NSEW-NSEW	10	42	253	11	23	rectangular	2.09	4.2	0.040	16.495
	NS-NSEW	10	42	253	11	23	rectangular	2.09	4.2	0.040	16.076
C	N-SEW	6	38	189	9	21	rectangular	2.33	6.3	0.032	16.421
	EW-NS	6	38	189	9	21	rectangular	2.33	6.3	0.032	16.000
	NSEW-NSEW	6	38	189	9	21	rectangular	2.33	6.3	0.032	14.140
	NS-NSEW	6	38	189	9	21	rectangular	2.33	6.3	0.032	14.105
D	N-SEW	6	30	105	5	21	rectangular	4.20	5.0	0.057	13.378
	EW-NS	6	30	105	5	21	rectangular	4.20	5.0	0.057	14.000
	NSEW-NSEW	6	30	105	5	21	rectangular	4.20	5.0	0.057	11.911
	NS-NSEW	6	30	105	5	21	rectangular	4.20	5.0	0.057	11.600
E	N-SEW	4	22	121	11	11	quadratic	1.00	5.5	0.033	13.182
	EW-NS	4	22	121	11	11	quadratic	1.00	5.5	0.033	12.000
	NSEW-NSEW	4	22	121	11	11	quadratic	1.00	5.5	0.033	11.045
	NS-NSEW	4	22	121	11	11	quadratic	1.00	5.5	0.033	10.909

Table B.1: Overview of layout population of section 6.4

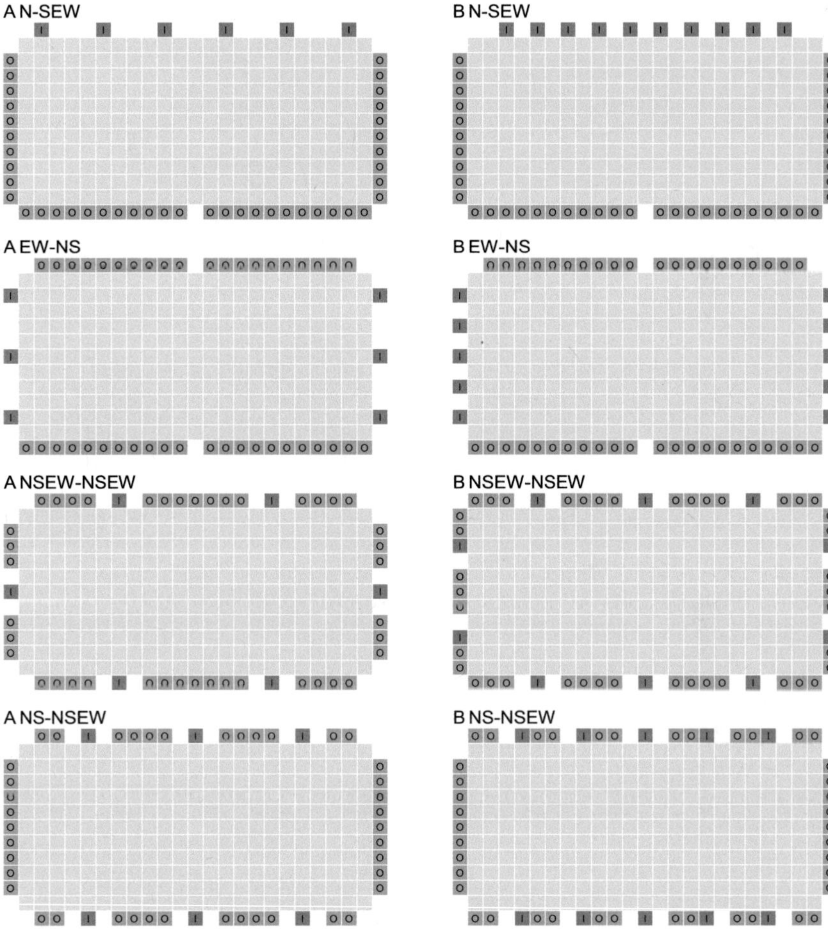

Figure B.1: Graphical representation of layout group A and B of section 6.4

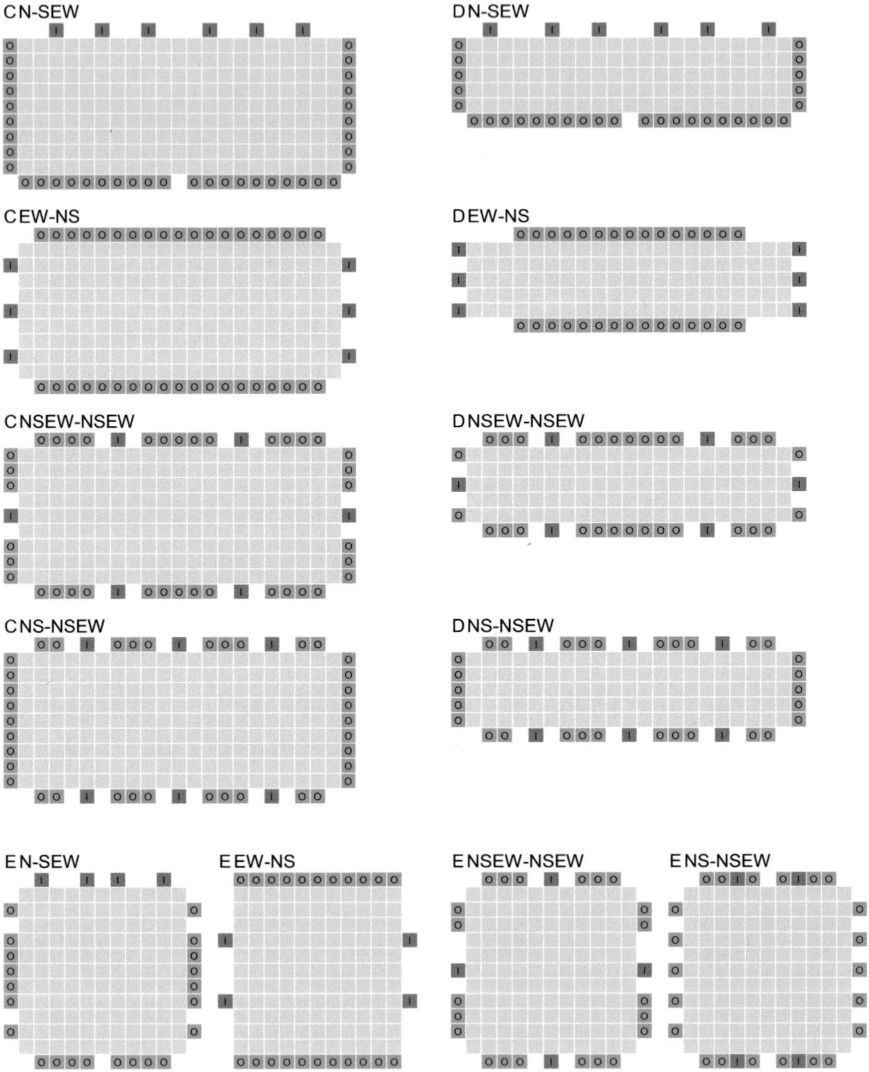

Figure B.2: Graphical representation of layout group C, D and E of section 6.4

Layout ID	# sources	# destinations	# modules	# rows	# columns	Shape	Aspect ratio	Source-destination-ratio	Source-module-ratio	Position of sources and destinations	Distance from sources to destinations	Sources close to each other
0	10	57	400	20	20	quadratic	1.00	5.7	0.025	N-ESW	22.833	
1	10	57	400	16	25	rectangular	1.56	5.7	0.025	N-ESW	22.842	
2	8	43	252	8	33	cropped	4.13	5.4	0.032	N-ESW	19.558	
3	8	43	253	11	23	rectangular	2.09	5.4	0.032	N-ESW	18.593	
4	8	43	252	7	36	rectangular	5.14	5.4	0.032	N-ESW	20.337	
5	7	40	195	13	15	rectangular	1.15	5.7	0.036	N-ESW	16.364	
6	4	20	195	13	15	rectangular	1.15	5.0	0.021	N-ESW	16.550	
7	7	40	200	10	20	rectangular	2.00	5.7	0.035	N-ESW	16.771	
8	7	34	200	10	22	cropped	2.20	4.9	0.035	N-ESW	16.592	
9	20	40	200	10	20	rectangular	2.00	2.0	0.100	N-ESW	16.825	x
10	7	40	200	10	20	rectangular	2.00	5.7	0.035	NSEW-NSEW	14.061	
11	7	40	200	10	20	rectangular	2.00	5.7	0.035	NS-NSEW	13.318	
12	7	40	200	10	20	rectangular	2.00	5.7	0.035	EW-NS	16.000	
13	7	40	200	5	40	rectangular	8.00	5.7	0.035	N-ESW	19.300	
14	7	40	198	6	33	rectangular	5.50	5.7	0.035	N-ESW	18.786	
15	7	45	198	6	33	rectangular	5.50	6.4	0.035	N-ESW	18.867	
16	6	35	150	10	15	rectangular	1.50	5.8	0.040	N-ESW	14.533	
17	12	35	150	10	15	rectangular	1.50	2.9	0.080	N-ESW	14.462	x
18	12	35	150	10	15	rectangular	1.50	2.9	0.080	NSEW-NSEW	12.443	
19	6	18	150	10	15	rectangular	1.50	3.0	0.040	N-ESW	14.370	
20	6	35	150	15	10	rectangular	1.50	5.8	0.040	N-ESW	14.567	x
21	6	35	150	15	10	rectangular	1.50	5.8	0.040	N-ESW	16.386	
22	6	35	150	6	27	cropped	4.50	5.8	0.040	N-ESW	15.190	
23	12	35	150	6	27	cropped	4.50	2.9	0.080	N-ESW	15.219	
24	12	35	150	6	27	cropped	4.50	2.9	0.080	NSEW-NSEW	14.307	
25	5	31	110	10	11	rectangular	1.10	6.2	0.045	N-ESW	12.548	
26	4	31	110	10	11	rectangular	1.10	7.8	0.036	N-ESW	12.613	
27	3	31	110	10	11	rectangular	1.10	10.3	0.027	N-ESW	12.634	
28	3	16	110	10	11	rectangular	1.10	5.3	0.027	N-ESW	13.083	
29	6	31	110	10	11	rectangular	1.10	5.2	0.055	NS-NSEW	10.726	

Table B.2: Overview of big layout population (ID 0 to 29)

Layout ID	# sources	# destinations	# modules	# rows	# columns	Shape	Aspect ratio	Source-destination-ratio	Source-module-ratio	Position of sources and destinations	Distance from sources to destinations	Sources close to each other
30	10	31	110	5	22	rectangular	4.40	3.1	0.091	N-ESW	13.694	
31	6	31	110	5	22	rectangular	4.40	5.2	0.055	N-ESW	13.780	
32	10	31	110	5	22	rectangular	4.40	3.1	0.091	NSEW-NSEW	12.229	
33	6	31	109	5	23	cropped	4.60	5.2	0.055	N-ESW	13.366	
34	4	20	50	5	10	rectangular	2.00	5.0	0.080	N-ESW	8.850	
35	4	20	50	5	10	rectangular	2.00	5.0	0.080	NSEW-NSEW	7.625	
36	4	20	50	5	10	rectangular	2.00	5.0	0.080	NS-NSEW	7.400	
37	4	20	50	5	10	rectangular	2.00	5.0	0.080	EW-NS	8.500	
38	4	19	49	7	7	quadratic	1.00	4.8	0.082	N-ESW	9.211	
39	4	20	51	3	17	rectangular	5.67	5.0	0.078	N-ESW	9.700	
40	4	17	51	7	9	cropped	1.29	4.3	0.078	N-ESW	8.529	
41	1	15	25	5	5	quadratic	1.00	15.0	0.040	N-ESW	6.400	
42	3	15	25	5	5	quadratic	1.00	5.0	0.120	N-ESW	6.444	
43	2	15	25	5	5	quadratic	1.00	7.5	0.080	N-ESW	6.467	
44	2	20	50	5	10	rectangular	2.00	10.0	0.040	N-ESW	8.600	
45	1	21	49	7	7	quadratic	1.00	21.0	0.020	N-ESW	8.571	
46	2	19	45	5	9	rectangular	1.80	9.5	0.044	N-ESW	8.316	
47	9	50	300	15	20	rectangular	1.33	5.6	0.030	N-ESW	20.122	
48	9	50	306	17	18	rectangular	1.06	5.6	0.029	N-ESW	20.064	
49	12	65	500	20	25	rectangular	1.25	5.4	0.024	N-ESW	25.672	
50	12	65	506	22	23	rectangular	1.05	5.4	0.024	N-ESW	26.410	
51	12	65	500	10	50	rectangular	5.00	5.4	0.024	N-ESW	28.546	
52	5	31	110	5	22	rectangular	4.40	6.2	0.045	N-ESW	13.490	
53	5	30	108	9	12	rectangular	1.33	6.0	0.046	N-ESW	12.447	
54	12	57	500	20	28	cropped	1.40	4.8	0.024	N-ESW	24.152	
55	10	54	400	16	26	cropped	1.63	5.4	0.025	N-ESW	22.130	
56	10	57	400	10	40	rectangular	4.00	5.7	0.025	N-ESW	25.093	
57	9	48	300	10	32	cropped	3.20	5.3	0.030	N-ESW	20.056	
58	9	50	300	10	30	rectangular	3.00	5.6	0.030	N-ESW	20.833	
59	4	20	270	18	15	rectangular	0.83	5.0	0.015	N-ESW	20.650	

Table B.3: Overview of big layout population (ID 30 to 59)

Layout ID	# sources	# destinations	# modules	# rows	# columns	Shape	Aspect ratio	Source-destination-ratio	Source-module-ratio	Position of sources and destinations	Distance from sources to destinations	Sources close to each other
60	6	18	210	14	15	rectangular	1.07	3.0	0.029	N-ESW	17.074	
61	4	18	40	5	8	rectangular	1.60	4.5	0.100	N-ESW	8.111	
62	12	35	150	10	15	rectangular	1.50	2.9	0.080	NS-NSEW	12.548	
63	12	34	153	9	17	rectangular	1.89	2.9	0.078	EW-NS	14.000	x
64	12	35	150	6	27	cropped	4.50	2.9	0.080	NS-NSEW	12.700	
65	12	35	150	6	27	cropped	4.50	2.9	0.080	EW-NS	16.500	x
66	10	32	110	5	22	rectangular	4.40	3.2	0.091	NS-NSEW	11.875	
67	10	32	110	5	22	rectangular	4.40	3.2	0.091	EW-NS	14.500	x
68	4	16	40	5	8	rectangular	1.60	4.0	0.100	EW-NS	7.500	
69	4	20	35	5	7	rectangular	1.40	5.0	0.114	NSEW-NSEW	6.600	
70	4	18	40	5	8	rectangular	1.60	4.5	0.100	N-ESW	7.944	
71	2	18	40	5	8	rectangular	1.60	9.0	0.050	N-ESW	7.833	
72	4	20	57	3	19	rectangular	5.67	5.0	0.070	N-ESW	10.300	
73	5	31	92	4	23	rectangular	5.75	6.2	0.054	N-ESW	12.742	
74	7	40	204	6	34	rectangular	5.67	5.7	0.034	N-ESW	18.379	
75	9	50	280	7	40	rectangular	5.71	5.6	0.032	N-ESW	21.971	
76	10	57	368	8	46	rectangular	5.75	5.7	0.027	N-ESW	25.468	
77	11	63	459	9	51	rectangular	5.67	5.7	0.024	N-ESW	27.694	
78	19	59	390	10	39	rectangular	3.90	3.1	0.049	N-ESW	24.390	
79	36	40	400	20	20	quadratic	1.00	1.1	0.090	EW-NS	21.000	x
80	19	53	273	7	39	rectangular	5.57	2.8	0.070	N-ESW	21.660	
81	28	52	500	20	28	cropped	1.40	1.9	0.056	N-ESW	25.327	x
82	24	48	407	11	37	rectangular	3.36	2.0	0.059	NSEW-NSEW	21.535	
83	16	60	361	19	19	quadratic	1.00	3.8	0.044	NSEW-NSEW	18.500	
84	18	58	407	11	37	rectangular	3.36	3.2	0.044	NS-NSEW	21.253	

Table B.4: Overview of big layout population (ID 60 to 84)

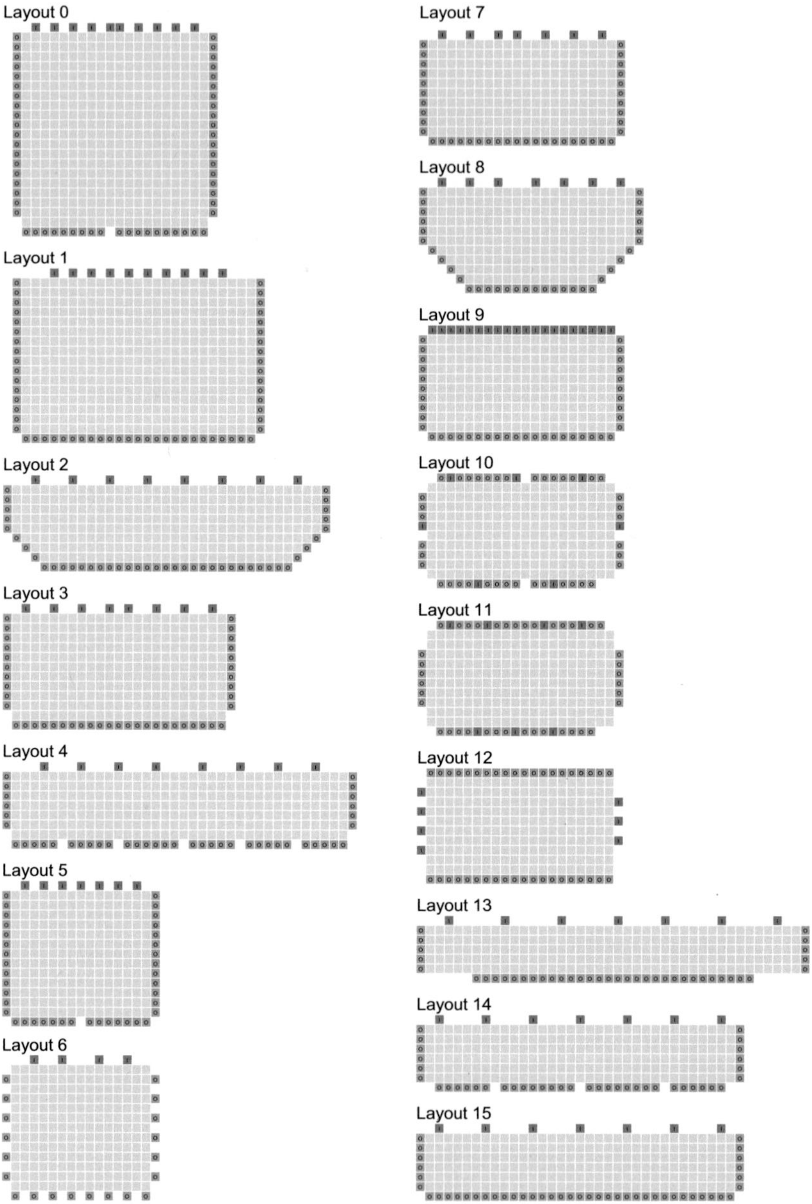

Figure B.3: Graphical representation of layouts with ID 0 to 15

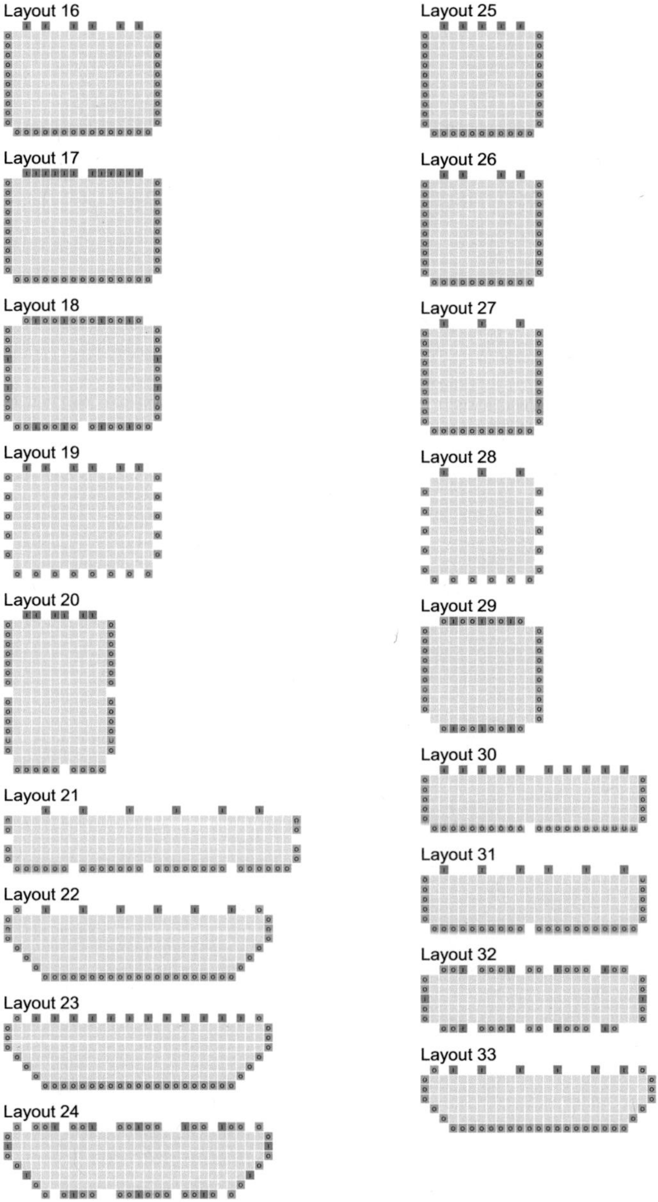

Figure B.4: Graphical representation of layouts with ID 16 to 33

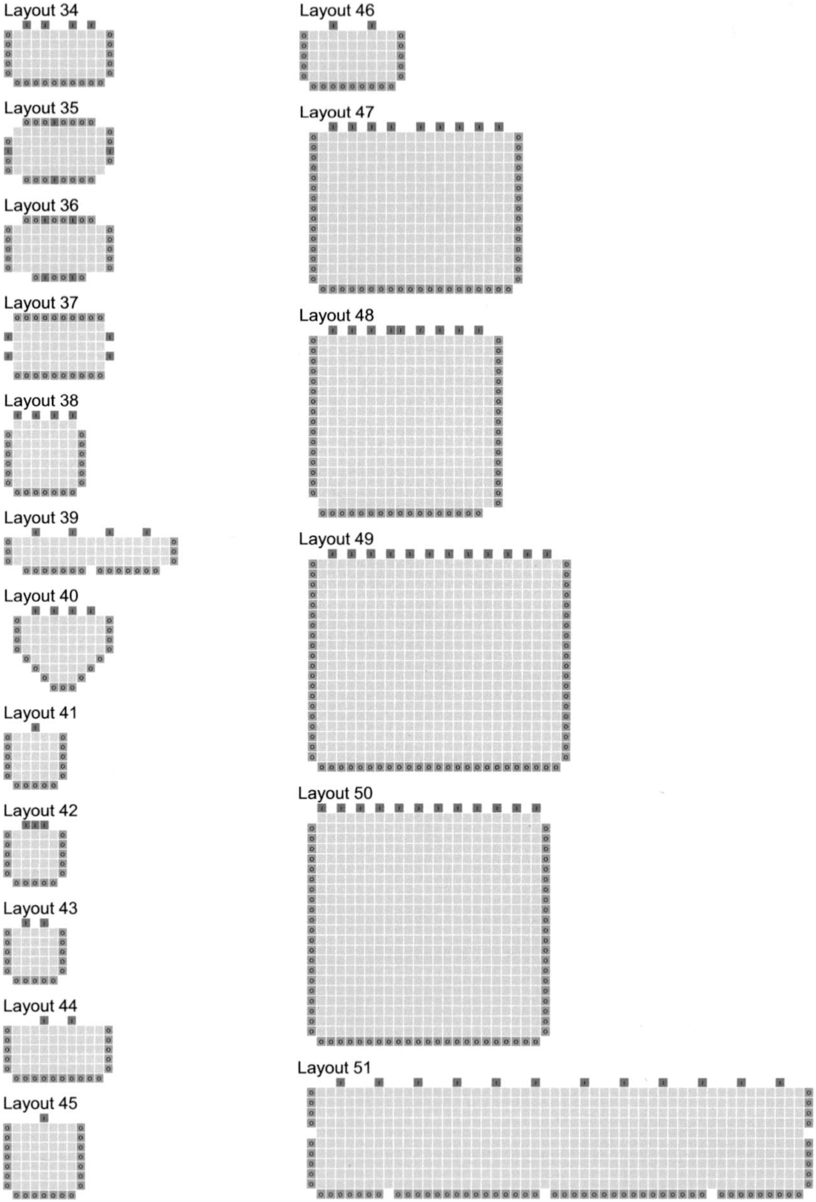

Figure B.5: Graphical representation of layouts with ID 34 to 51

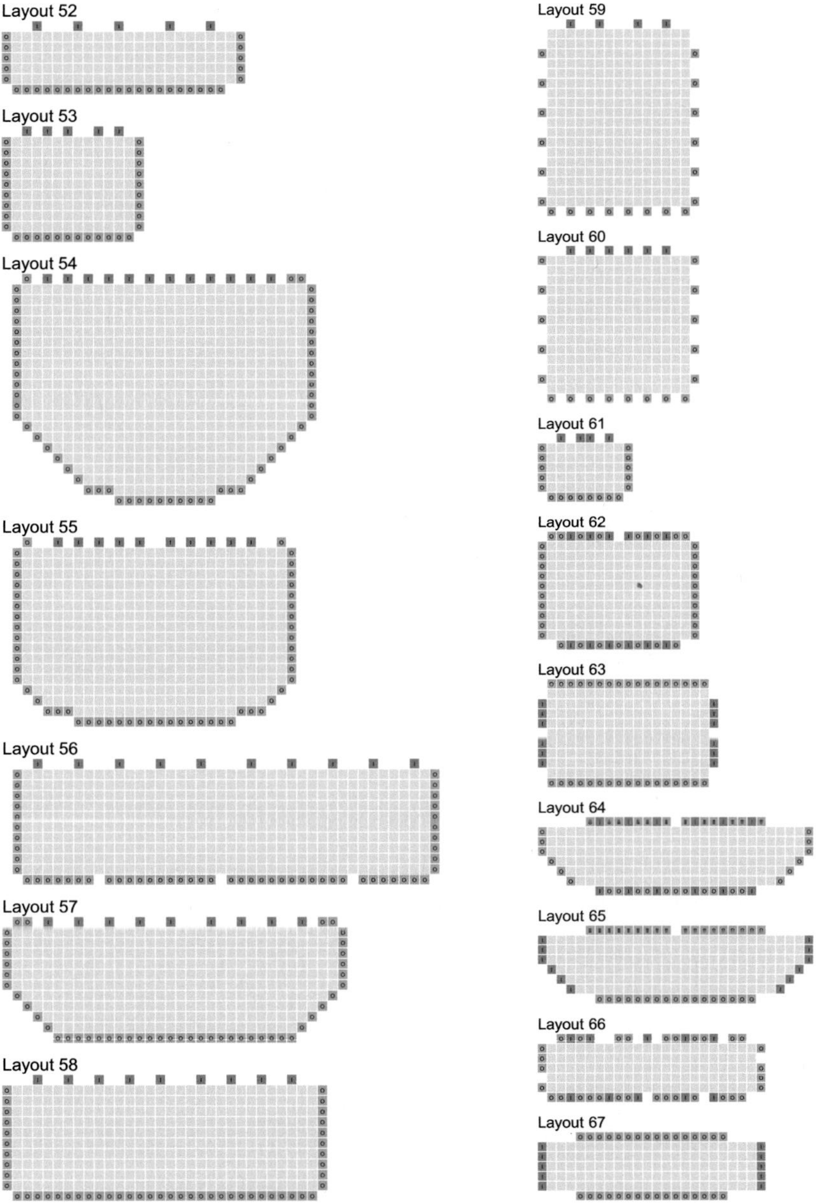

Figure B.6: Graphical representation of layouts with ID 52 to 67

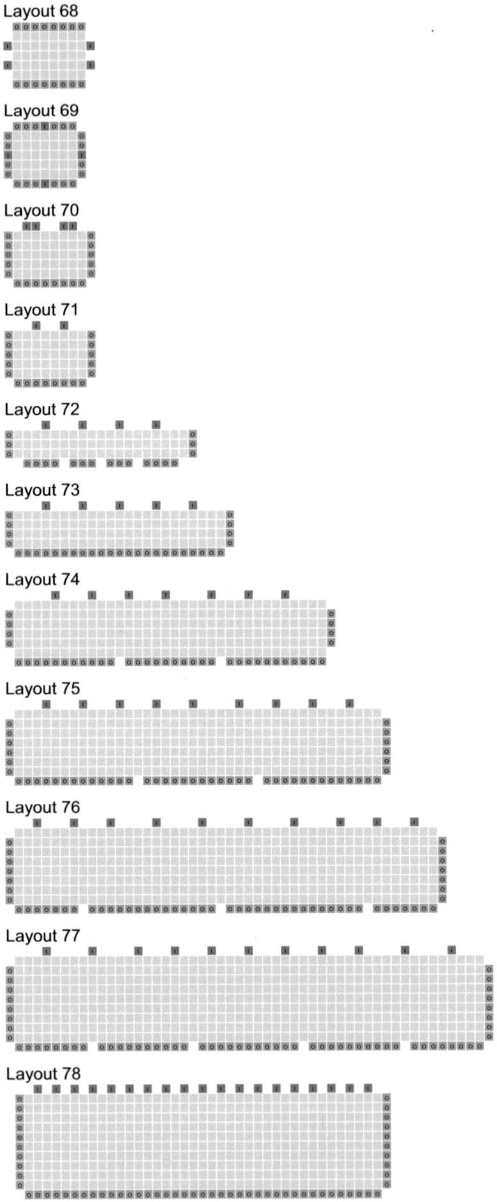

Figure B.7: Graphical representation of layouts with ID 68 to 78

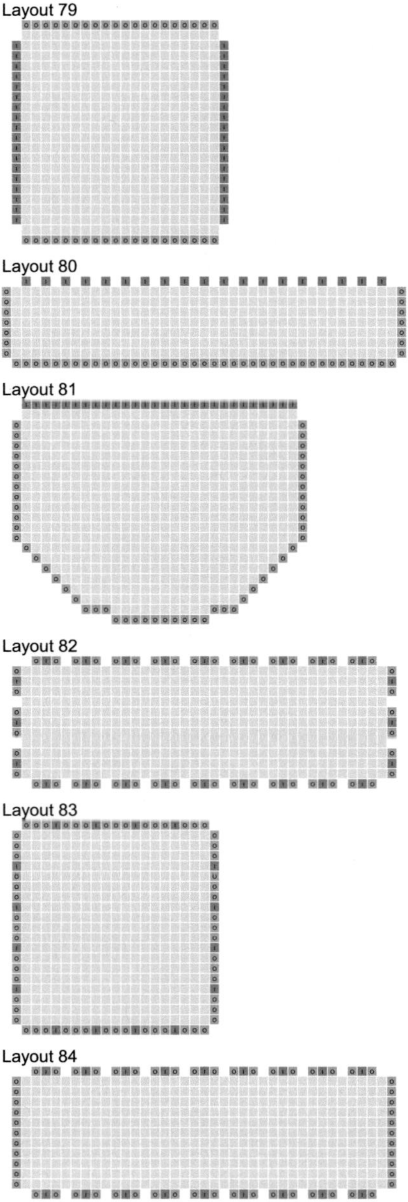

Figure B.8: Graphical representation of layouts with ID 79 to 84

C Results for Partial Route Reservation

Message count limit	Maximal Remaining Distance to Destination	Throughput as portion of maximally achieved throughput		F	p-value
		μ	σ^2		
100	0.75	99.04%	0.04%	6.39	0.0000
300	0.75	98.91%	0.01%		
500	0.75	98.80%	0.01%		
100	0.5	99.64%	0.00%		
300	0.5	99.00%	0.01%		
500	0.5	98.79%	0.01%		
100	0.25	99.19%	0.01%		
300	0.25	98.77%	0.01%		
500	0.25	98.70%	0.01%		
100	all	99.29%	0.02%	16.62	0.0000
300	all	90.09%	0.01%		
500	all	98.76%	0.01%		
all	0.75	98.91%	0.02%	4.34	0.0134
all	0.5	99.14%	0.01%		
all	0.25	98.89%	0.01%		

Figure C.1: Results of analysis of variance for parameters of partial route reservation

D Students' Thesis related to GridSorter

Many different students supported my work on GridSorter. I very much appreciated the close collaboration, fruitful discussions leading to good ideas and helpful results. Some of them are cited in my PhD thesis, but I also want to thank the others:

Florian Geiser Sortieren mit Plug-and-Play Fördertechnik *Diplomarbeit* (2012)

Jörg Schmidtobreick Herausforderungen an Routingprozessen bei dezentral gesteuerten Materialflusssystemen *Seminararbeit* (2012)

Thomas Lienert Darstellung, Analyse und Optimierung des Flex-Förderer-Reservierungsprozesses *Seminararbeit* (2013)

Vanessa Market und Anja Fechtig Stand der Technik und zukünftige Lösungsansätze bei der Warensortierung *Seminararbeit* (2013)

Emilie Vannerot Das Pull-Prinzip in der dezentralen Steuerung *Masterarbeit* (2013)

Fernando Lauckner Das Twitter-Prinzip in der Steuerung eines Plug&-Play-fähigen Sorters *Diplomarbeit* (2013)

Carlos Corrales Control algorithms for a Plug&Play sorter with multi-directional conveying *Master Thesis* (2013)

Johannes Schmitt Modellierung eines dezentral gesteuerten Sorters mit Hilfe der Graphentheorie *Diplomarbeit* (2013)

Benedikt Fuß Darstellung, Analyse und Optimierung des FlexFörderer-Deadlock-Handling-Prozesses *Seminararbeit* (2013)

Theresa Voss und Katja Rabenseifner Literaturrecherche zur Verkehrsaufkommengestützten Navigation und die Übertragung in die dezentral gesteuerte Warensortierung *Seminararbeit* (2013)

Thomas Lienert Algorithmenentwicklung für einen dezentral gesteuerten Sorter mit multidirektionalem Förderantrieb *Masterarbeit* (2013)

Ernst Viktor Prohl Binäres Optimierungsmodell für den GridSorter *Hilfswissenschaftlicher Mitarbeiter* (2014)

Theresa Gattermann Layoutanalyse bei einem dezentral gesteuerten Sorter *Bachelorarbeit* (2014)

Marius Thoma und Martin Rometsch Sortieralgorithmen in der Spieletheorie und der Informatik mit Bezug auf einen dezentral gesteuerten Plug&Play-Sorter *Seminararbeit* (2014)

Lisa Prössdorf Erweiterung eines dezentral gesteuerten Sorters um die Funktion des Speicherns *Masterarbeit* (2014)

Dominik Colling Steuerung eines Plug&Play-Sorters für unterschiedliche Ladungsträgergrößen *Masterarbeit* (2014)

Almut Demel Leistungsuntersuchung für Einsatzfälle des GridSorters *Hilfswissenschaftliche Mitarbeiterin* (2014)

Immanuel Ketterer Analyse eines zeitfensterbasierenden Reservierungsalgorithmus eines Plug&Play Sorters *Seminararbeit* (2014)

Alexandre Araujo Analysis of possible applications of the GridSorter in a packaging handling facility *Bachelor Thesis* (2014)

Lisa Wirsig Analyse der Einsatzmöglichkeiten des GridSorters in der Lagervorzone *Bachelorarbeit* (2014)

Justine Portier Untersuchung von Einsatzmöglichkeiten des GridSorters in der Kommissionierung *Bachelorarbeit* (2014)

Linda Lehmann Erweiterung der Steuerung eines modularen Sorters um die Funktion der Sequenzierung *Bachelorarbeit* (2014)

Alexander Werling Weiterentwicklung des dezentral gesteuerten Routings eines modularen Sorters zur Effizienzsteigerung auf großen Layouts *Bachelorarbeit* (2015)

Janine Gieringer, Franziska Ötjen und Johannes Schuh Literaturrecherche rund um den GridSorter *Seminararbeit* (2015)